T0324900

The Scientific Exploration of Venus

Venus is the nearest planet to the Earth, observed since ancient times as the beautiful, brilliant Morning or Evening 'Star' in the night sky. Venus is also the world most similar to ours in size, mass and composition. Before the space age began, it was widely expected that conditions on the surface of our neighbour would resemble a more tropical version of the Earth. In fact, recent robotic missions to the planet have revealed a hot, dry climate with a dense carbon dioxide atmosphere and clouds rich in sulphuric acid. There are no seas; the surface is dominated by thousands of volcanoes, and it lacks a protective magnetic field to shield it from energetic solar particles and cosmic rays.

In this book, a leading researcher of Venus addresses these contrasts while explaining what we know through our investigations of the planet. Venus presents an intriguing case study for planetary astronomers and atmospheric scientists, especially in light of the current challenges of global warming, which supports, and potentially threatens, life on Earth. Scientifically rigorous, yet written in a friendly non-technical style, this is a broad introduction for students, and astronomy and space enthusiasts.

FREDRIC W. TAYLOR is Emeritus Halley Professor of Physics at Oxford University. He is a senior figure in the planetary science community, and has been involved in NASA and ESA missions to study Mars, Venus, Jupiter and Saturn. He is also a prolific author; in addition to his *The Scientific Exploration of Mars* (2009), he has written *The Cambridge Photographic Guide to the Planets* (2001) and the textbooks *Elementary Climate Physics* (2005) and *Planetary Atmospheres* (2010), and he has co-authored five other books. His lead roles in the *Pioneer Venus* and *Venus Express* missions give him a unique and authoritative perspective of this area. He is the recipient of numerous awards including 13 NASA Achievement Awards; The Bates Medal of the European Geophysical Society for Excellence in the Planetary Sciences; and The Arthur C. Clarke Lifetime Achievement Award.

The Scientific Exploration of Venus

Fredric W. Taylor
University of Oxford

CAMBRIDGE
UNIVERSITY PRESS

CAMBRIDGE
UNIVERSITY PRESS

University Printing House, Cambridge CB2 8BS, United Kingdom

One Liberty Plaza, 20th Floor, New York, NY 10006, USA

477 Williamstown Road, Port Melbourne, VIC 3207, Australia

314-321, 3rd Floor, Plot 3, Splendor Forum, Jasola District Centre, New Delhi - 110025, India

79 Anson Road, #06-04/06, Singapore 079906

Cambridge University Press is part of the University of Cambridge.

It furthers the University's mission by disseminating knowledge in the pursuit of education, learning and research at the highest international levels of excellence.

www.cambridge.org
Information on this title: www.cambridge.org/9781107023482

First published 2014

A catalogue record for this publication is available from the British Library

Library of Congress Cataloging in Publication data
Taylor, F. W., author.
The scientific exploration of Venus / Fredric W. Taylor, University of Oxford.
  pages   cm
Includes bibliographical references.
ISBN 978-1-107-02348-2
1. Venus (Planet) – Exploration.   I. Title.
QB621.T38   2014
559.9′22–dc23   2013048108

ISBN 978-1-107-02348-2 Hardback

Contents

Contents

The plates are situated between pages 140 and 141.

Overview

Venus is well known to everyone as the brightest star in the evening or morning sky. Of course, this brilliant stellar object is not actually a star, but a planet, the closest to Earth and, it turns out, the one that most resembles our own world in size and composition. It should therefore be the easiest to explore by astronomers observing from the Earth with telescopes, and indeed there is a history of Venus observations that extends back to the earliest recorded times. However, as observations got better with the invention and improvement of the telescope, the result was often frustration because so little detail could be seen on our bright neighbour. Instead it was found that the surface is shrouded, apparently at all times and at all places, by thick layers of nearly featureless cloud. It was not until the first spacecraft arrived, just half a century ago, that the true character of Venus began to be revealed.

This book presents an account of the exploration of Venus, from the earliest days to the latest research using planetary space missions. It also ventures some visions of the distant future when Venus is explored by humans, and might once again have an Earthlike climate (if indeed it once did in the past, as many scientists believe). The space projects and other types of investigation are covered in some detail, especially their scientific objectives and accomplishments. As in the author's recent book, *The Scientific Exploration of Mars* (Cambridge University Press, 2009), the aim is to be scientifically rigorous but at the same time understandable by non-experts, such as amateur astronomers, students and interested people from all walks of life. Fifty years of experience in talking to special interest groups, schools and colleges, literary festivals, informal gatherings and the media have shown that there is wide interest in penetrating the jargon and protocol of scientific research on the nearby planets and what follows attempts to address this need.

The chapters are organised in sections which deal first with the accumulation of our present knowledge, then with the key problems remaining and the research currently under way to look for answers and an understanding of how the planet evolved, how it resembles the Earth and how and why it differs. Inevitably, the focus is on space missions, from which most of our modern insights have come. The approach here is different from any of the (relatively few) existing books about Venus, with a harder core that centres on an in-depth appreciation of the science and mission architecture and activities, while maintaining a format and style that should not put off the more general reader.

Prologue

Venus is the closest planet to the Earth and, with the obvious exceptions of the Sun and Moon, it is easily the brightest object which regularly appears in our sky. Near inferior conjunction, the time of closest approach when the two planets are separated by a mere 40 million kilometres, a modest telescope will reveal Venus as a brilliant, featureless crescent. The crescent-shaped appearance is characteristic of a body that is closer to the Sun than the observer, as Galileo realised when he turned his primitive telescope on Venus four centuries ago.

Modern values for the basic physical characteristics of the planet are very close to Earth's, with Venus having about 82 per cent of the mass, 95 per cent of the diameter, and about the same density, suggesting a similar internal composition and structure of a rocky mantle with an iron-nickel core. No other planet matches ours so closely: Mars, for instance, has only about a tenth of Earth's mass, and just over half the diameter. Mercury is even smaller, weighing in at only 5 per cent of the mass of Earth, and the Moon at just over 1 per cent. So the Earth's nearest neighbour is also its only real twin in our solar system.

Despite its clear resemblance to our home planet, Venus has never been as popular with authors of books as our other planetary neighbour, Mars. This applies whether they are scientific or popular, factual or science fiction works, and is despite the fact that Mars is small compared with Earth or Venus, and significantly farther away. Lying outside Earth's orbit, Mars approaches on average to a distance of about 77 million kilometres, whereas at just over half that, Venus is only a proverbial stone's throw away for today's rockets and their scientific payloads. Even manned missions to Venus could readily be contemplated with no better technology than that which we already have, if only the destination were more appealing.

Until recently, the main reason why Venus has lost to Mars in the popularity stakes was that a thick, permanent veil of cloud hides the Venusian surface. Therefore, it was much harder for authors to describe or visualise the landscape and any associated weather and seasons. It was also harder to contemplate landing, exploring and living there. It was not that Venus did not seem a promising abode for life, in those early years before the space age, but that so much was left to the imagination as to what the conditions were under

Prologue

which life might exist. This did not stop speculative fiction writers like C.S. Lewis, Edgar Rice Burroughs and Frank Hampson from indulging their fantasies and entertaining us with tales of adventures on an Earthlike Venus.[1]

But now that we know the truth about the hellish environment on Venus, the chances of finding life there have faded, leaving cold, almost airless Mars looking a much better prospect for expeditions to explore and perhaps discover biological artefacts, alive or dead. It is almost impossible (although we shall try, in one of the final chapters) to imagine manned landings on a surface hot enough to melt lead and with the sort of pressures that on Earth we associate with the deep ocean bed. Even science fiction writers prefer not to fly in the face of known facts when framing their stories, at least not too many of them or too blatantly.

So Venus is mostly out of favour as a setting for stories and movies, as well as for new scientific space missions, losing out regularly to Mars in particular. But many of us Earthlings still wish to understand what Venus can tell us about our origins as part of the planetary system that is home to both worlds. Despite the conditions there, could Venus possibly host some form of life? And most of all perhaps, Venus gets interesting again, even in a life-supporting sense, once we realise that the very reason that it is now so inhospitable, hot and dry, is the same as the cause for most of our current apprehension about changes to the environment on the Earth. The greenhouse effect, fuelled by carbon dioxide, reigns on Venus as it does here, and when it gets out of control things get very tough for most of the familiar life forms, including humans.

The enduring scientific interest in Venus's climate, including the urge to explore and the handy proximity of Venus for relatively cheap missions, has been enough to lead to the attempted dispatch of no fewer than 44 spacecraft from Earth since interplanetary spaceflight began in 1963. About half of them were successful (see Appendix B). Until recently the leading players have been the Soviet – now Russian – and American space agencies, but recently we have seen the first flights to the planet from Europe (with *Venus Express*) and Japan (with *Akatsuki*), and longer-term plans and speculations from all of them. The chapters that follow tell the story of how the missions and other initiatives came about, what they did and what we have learned.

The narrative is in three parts. First, we cover the pre–space age of Venus exploration, from the earliest times, telling why experts used to expect dinosaur-infested swamps before they began to get hints that things there were not so benign. The chapters on space-age exploration deal with each mission and major breakthrough individually. The central section summarises current knowledge by scientific topic – surface geology, atmospheric composition, weather, climate and so on – and highlights the remaining mysteries. In the third and final part we will see what plans exist for likely future visits to Venus, all the way to the prospects for eventual human exploration and the long-term evolution of the planet itself.

---

[1] In 1950 Hampson created for the *Eagle*, a weekly paper aimed at teenaged boys, the first episode of what became a long-running serial chronicling the exploits of Dan Dare, Pilot of the Future, a sophisticated and imaginative comic strip that was initially set on Venus.

xiv

The main themes are exploration – what has been discovered, and how it was done – and science, why Venus is like it is and what we learn from studying it. Comparisons with the Earth are unavoidable and indeed essential at every step. The approach is to seek to be as complete and rigorous as possible about the science, and the engineering challenges addressed by the different spacecraft and instruments, without being so technical that the only readers who understand are the specialists. Instead the book should be useful to students, as an introduction and overview, but is mostly styled so that laypersons and amateurs interested in the planets will find it an informative and enjoyable read as well.

In order to improve accessibility for those without a scientific background I have tried to avoid, or at least simplify, the jargon which is used by the professionals, and add lots of explanatory notes. Details that are not essential for a *general* appreciation of the study of Venus, such as the quantitative error and uncertainty limits on individual measurements, are deliberately glossed over.

Anyone who finds they want to look a little deeper into the scientific topics, including basic definitions and equations, could start with the textbooks I wrote for the Oxford Physics undergraduates, *Planetary Atmospheres* (2010) and *Elementary Climate Physics* (2005).

A note on scientific units

Popular units for physical quantities are used wherever possible in preference to the official conventions used in the formal literature. This is intended to improve readability and to emphasise that this is a general and not a rigorous description of the scientific exploration of Venus. For the same reason, to provide a read that is as user-friendly as possible, technical terms, detailed numeracy and quantification are all avoided except where absolutely essential, and then they are minimised and explained.

Temperatures are expressed in the familiar centigrade (Celsius) scale (°C). Where temperatures are mentioned in degrees, degrees centigrade are to be understood. Occasionally, it is better to use the absolute scale in kelvins (K). This is the same as centigrade in that a temperature difference of 1K = 1°C, but starts at absolute zero, so that 273.15K = 0°C.

Pressure is usually given in atmospheres, where one atmosphere is the mean surface pressure on the Earth, except that very small pressures are in millibars (mb), one thousandth of an atmosphere, bars and atmospheres being essentially identical units for present purposes and 'milliatmosphere' is not used.[1]

Distances are in kilometres, and if very small (such as wavelengths and the diameters of cloud particles, for example) the unit used is the micron (μm), the common abbreviation for a micrometre, one-millionth of a metre.

Mass is usually in kilograms, but very large values are expressed in tons, which means metric tons, equal to 1,000 kilograms (sometimes called 'tonnes', but not here). Small masses are in milligrams (mg) or occasionally micrograms (μg).

---

[1] A bar is actually 0.987 atmospheres. The reason two such similar units exist is because the bar was introduced (by the Meteorological Office in 1909) to replace the older unit with one that was exactly 100,000 pascals (Pa). Pressures in Earth's atmosphere other than at the surface are of course normally less than a bar and the millibar (a thousandth of a bar) is a commonly used unit, although officially this has been replaced by the hectopascal (1hPa = 1 mbar) to achieve standardisation.

Acknowledgements

Thousands of people contributed to the work described in the following pages, as planetary scientists and experimenters like the author, or as the politicians, mission planners, managers and engineers who put their complex plans and projects into practice. It is impossible to mention everyone by name; indeed, I have tried to avoid using names as far as possible in order to maintain the top-level nature of the main narrative for its intended general audience. (A future book, *Exploring the Planets*, will take the opposite tack and focus on people and anecdotes.)

My own perspective, from my involvement in Venus studies from the early 1970s until the present day, informs the design and perspective of the book and any omissions, biases, or (hopefully few) errors it may have. Still, the effort and innovation of others made it all possible. I feel fortunate to have known and worked with many of them.

A few key individuals will be mentioned by name along the way when it becomes essential to do so, but the rest, just as important and just as vital to success, can for practical reasons receive only this blanket acknowledgement. There is a 'References and Acknowledgements' section at the end that indicates where the reader can find an original or more in-depth treatment of the topic of the chapter. Figures and quotes from other publications are acknowledged here, as well.

I am grateful to colleagues who have read the draft manuscript and made invaluable suggestions, especially Colin Wilson and Martin Airey. Special thanks are due to Dr D. J. Taylor, who prepared the artwork for the book, including drawing the original figures and re-drawing many others.

Part I
Views of Venus, from the beginning to the present day

Observations of Venus with the naked eye as a prominent planet or 'wandering star' were recorded by the Babylonians around 3000 BC, and have continued ever since. All the major civilisations have contributed knowledge and myth to a recondite and, until recently, quite abstruse concept of our nearest companion in space beyond the Moon. With the invention of the telescope in about 1610 it became clear to Galileo that Venus shone in reflected light from the Sun and had phases like the Moon, leading eventually to an understanding that Venus is not any kind of star, but an Earthlike object, one that orbits closer to the Sun than we do. The presence of an atmosphere on Venus, filled with cloud that veiled the entire planet at all times and prevented the observation of surface features, was recognised and refined from the 1760s onwards, and the principal composition of the atmosphere was established in the 1930s.

Most of what we now know about our planetary neighbour has come from observations by spacecraft that flew to the planet and collected data during flybys, from orbit, or in a few cases during descent to the surface on parachutes. This phase of exploration began in 1962 and has continued fitfully up to the present, including one spacecraft, *Venus Express*, still operating in orbit around the planet at the time of writing (February 2014). A complete list of missions to Venus is given in Appendix A, and the outline description of the planet that has been gleaned from these, and from ongoing observations with telescopes on the Earth, is summarised in Appendix B.

This first section of the book elaborates on these early studies and sets the scene for the discussion in Part II of the most recent work, including studies carried out by space missions to Venus from Europe and Japan, as well as the pioneering ventures from Soviet Russia and the United States. The progress made shows Venus is indeed Earthlike, but with many features that are curiously different and some that remain difficult to explain even after 5,000 years of observing and wondering.

Chapter 1
The dawn of Venus exploration

The Evening Star and the Morning Star

Everyone has seen Venus, as a bright, starlike apparition in the evening sky, following the Sun down towards the horizon and setting a few hours later. At various other times of the year, there comes a brief season where an early bird can see Venus rise brilliantly before the Sun, climbing higher until it seems to dim and vanish as the sky brightens after sunrise. When it rises before the Sun, people have long called Venus the Morning Star; half an orbit later, when on the other side of the Sun so that the Sun sets first, Venus is the Evening Star. Before Copernicus promoted the idea that planets orbit the Sun, it was not obvious that these two phenomena were the same body, and early civilisations had distinct names for them. To the Greeks, they were Phosphoros and Hesperos.

For much of the year, Venus sets and rises so near the Sun that we tend not to notice it. During the day, like the true stars at vastly greater distances, Venus is still overhead and just as bright, of course, but it is hard to see because the contrast with the dark sky is lost when the Sun is up. It can be studied during the day if a telescope is used to shut out most of the sunlight, and even with ordinary binoculars if you know where to look.[1] In any observations made over a period of a few months, Venus can be seen to exhibit lunarlike phases (Figure 1.1).

As viewed from the Earth, Venus traces a flattened 'figure eight' pattern with the Sun at the centre (Figure 1.2). Sometimes, but rarely, Venus travels across the disc of the Sun when at its closest to the Earth, or behind the Sun when at its farthest, and we witness a transit.

At closest approach to the Earth (inferior conjunction), not only is Venus near the Sun in the sky, making viewing difficult, but also the side facing us is dark (Figure 1.2). A fully illuminated disc can be seen, again with difficulty, only when Venus is on the far side of the Sun, at so-called superior conjunction.

---

[1] There are also reports of Venus seen with the naked eye during the day. The most famous of these involved Napoleon Bonaparte, whose attention was called to the phenomenon while he was delivering an open-air, midday address to a crowd in Luxembourg in 1796. A similar apparition was reported in Washington DC on the day of Abraham Lincoln's second inauguration in 1865.

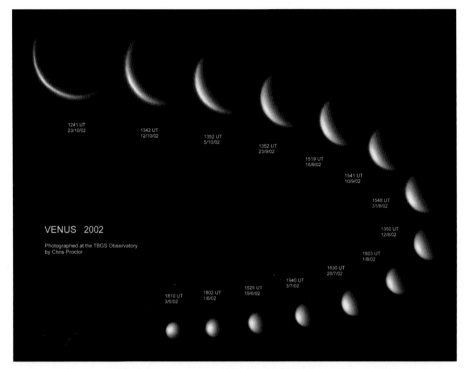

Figure 1.1 Venus at different points in its orbit, showing the phases from near superior conjunction (beyond the Sun, at bottom) to near inferior conjunction (closest to Earth, top).

The time between sunset and Venus's disappearance below the horizon, and vice versa, is greatest when it appears at its greatest separation from the Sun, called opposition, and viewing conditions are usually best then. However, the time when Venus appears brightest to us is not in fact exactly at opposition, but halfway between opposition and inferior conjunction, when the trade-off between size of the disc and the portion illuminated, what astronomers call the phase, is optimum.

Because it is at times such a brilliant object, Venus has been observed since the earliest times and used as an object of veneration[2] and as a celestial calendar by many early civilisations, most notably the Mayans in South America. There has been some interesting debate as to whether pre-telescopic observers could see the crescent shape of Venus with the naked eye. Written references to 'horned Venus' and implications that some symbolic crescents in art and heraldry might relate to Venus rather than the Moon would seem to support this idea, but the difficulty we have in achieving the feat today suggests otherwise (Plate 2).

The angle between a line from the observer on Earth to the centre of the Sun, and the corresponding line to Venus, is never more than 47.5 degrees (Figure 1.2). So, when the Sun is just set and our eyes are shielded from most of its light, allowing Venus to appear

---

[2] Nowadays of course most frequently associated with the Roman goddess of love, from whom the planet takes its modern name (Plate 1).

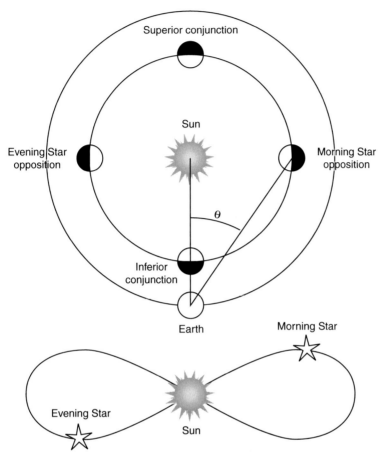

Figure 1.2 The orbits of Venus and Earth seen from above (top), and the apparent motion of Venus in the sky as viewed from Earth (below). It appears brightest at the points marked by stars, while approaching maximum elongation (i.e. the greatest apparent distance from the Sun). At that time, through a telescope it has the appearance of a crescent, as illustrated in Figure 1.1.

brightly, the planet is never more than halfway between the horizon and the zenith, that is, its elevation is always in the lower half of the sky. Mercury rises only half as far as Venus. The earliest recorded interpretation of these easily and much observed facts as meaning that Venus and Mercury orbit the Sun, and not the Earth, was made by Heraclides around 350 BC. Also in Greece, around 70 years later, Aristarchus surmised that the Earth, and everything else in the Universe, did the same.

The idea that the Earth was not the centre of the universe was too radical in those days, especially for the clerics, and of course was reluctantly accepted only much more recently. The breakthrough came in the early 1600s when the arrival of even very crude telescopes soon had Galileo following Copernicus and proclaiming the phases of Venus as a Moonlike phenomenon ('Cynthiae figuras aemulatur mater amorum') and making the heretical deduction that Venus must orbit the Sun. Not only that, but the 'Morning and then Evening' Star behaviour and the crescent phases must occur because Venus, and the much-dimmer and less well-observed Mercury, orbit closer to the Sun than the Earth. The other planets, which behave quite differently in the sky, lie outside the Earth's orbit.

In particular this must be true of the fourth member of the inner planet family, Mars. Mars is our second-nearest planetary neighbour, but Venus is larger (nearly as large as the Earth, see Plate 3) and significantly closer. Venus also has more cloud cover than the other three inner planets, not excluding the Earth, which makes its visible surface more reflective.[3] These three factors – size, proximity and albedo – explain why Venus can appear so bright.

Once it became an accepted principle for heavenly bodies to orbit the Sun and each other, it also became logical to wonder whether Venus had any satellites. It has been pointed out that if Venus had a moon on the same scale as the Earth's, it would be easily visible to the naked-eye observer here, including, of course, the ancients.[4] Under optimum conditions, the Venusian moon as seen from the Earth would be separated by more than a solar diameter from its parent in the night sky, and would be as bright as Saturn. Asimov points out that this obvious demonstration of one planetary-sized object orbiting another would have had a profound effect on the philosophers who pondered the nature of the universe, comparable to that after Galileo's observation of four large moons of Jupiter following the invention of the telescope. The Copernican revolution might have come thousands of years sooner, and Galileo might have been spared persecution, amongst many other consequences.[5]

Transits: Venus crosses the disc of the Sun, but rarely

If Earth and Venus orbited in exactly the same plane, we would see the disc of Venus as a dark spot crossing the Sun (a 'transit') every time the planet reached inferior conjunction, that is every 1.6 years, and then passing behind it 288 days later. However, the two orbital planes are tilted with respect to each other at an angle of 3.4 degrees, and this, plus the timing of the alignments of the three bodies in a straight line, means that transits are actually quite infrequent phenomena. Most of the time, Venus passes above or below the solar disc as seen from Earth.

About once a century, however, the path traced by Venus does cross the Sun, and then it does so twice in eight years, once in the upper and once in the lower solar hemisphere, as it moves from above the Sun to below, or vice versa. The pattern repeats every 243 years, with pairs of transits 8 years apart separated by gaps of 121.5 years and 105.5 years, the most recent pairs being in June 2004 and June 2012 (Plate 4). Before that, they were in December 1874 and December 1882, while the next will not take place until December 2117 and December 2125.

---

[3] Cloudy Venus has an albedo (from the Greek meaning 'whiteness') of about 0.76, which means it reflects all but 24 per cent of the sunlight that falls on it. Mars, by contrast, has little cloud cover and most of the reflected radiation comes from the relatively dark, rocky surface. The result is an albedo of only about 0.2. Earth is somewhere in between, with partial cloud, ice and ocean cover which together deliver an albedo in the region of 0.3.

[4] By Isaac Asimov, for example, in *The Tragedy of the Moon* (London, 1975).

[5] By an unusually, for him, arcane argument, Asimov suggests that these consequences would also mean that mankind nowadays 'may well be approaching the end of its days as a technological society' (ibid., pp 15–26).

The reason for this somewhat bizarre pattern of events has to do with the orbital periods of both planets, as well as their inclinations. It takes 243 years for Earth and Venus to return to the same relative positions because Venus travels around the Sun exactly 395 times in that many Earth years. A second orbital resonance, also probably coincidental, has thirteen Venus orbits in almost exactly eight Earth years, giving the eight-year separation of transits when the line from Earth to Venus at inferior conjunction intersects the Sun.

Other factors, such as the eccentricity[6] and precession[7] of both orbits, complicate the calculation, so the task of determining the timing of transits accurately is best done on a computer. Johannes Kepler attempted some predictions as early as 1627, but although he got the year of the next Venus transit right (1631), he did not realise that it would not be visible in Europe, nor that another was to come eight years later that would be. Improved calculations by Jeremiah Horrocks led him to make observations of the transit in 1639, from which he estimated the size of Venus and computed the first modern value of the distance from the Earth to the Sun,[8] allowing a scale to be put on the rest of the Solar System using Kepler's laws of planetary motion.

Later observers, most famously those travelling with Captain Cook on his first voyage around the world at the time of the transit of 1769, used the method derived by Edmond Halley (Figure 1.3) to get an improved value for the astronomical unit by simultaneous measurements from widely separated baselines on Earth.[9] Cook's observations (Figure 1.4)

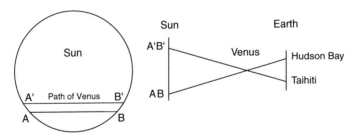

Figure 1.3 Halley's method for determining the Earth–Sun distance (1 AU) by observations of the transit of Venus across the solar disc from two widely separated locations on Earth. The observers measure the separation AB to A'B' as accurately as possible from the time taken for Venus to cross the solar disc in both cases, and use the fact that the Sun–Venus distance is 0.723 AU as determined from Venus's orbital period using Kepler's third law.

---

[6] The eccentricity of an orbit is a measure of its departure from a perfect circle, something which changes slowly over long periods of time. The current values for Earth and Venus are 0.0167 and 0.0068, while a circle is of course 0.0000.

[7] 'Precession' refers to how the alignments of the two non-circular orbits vary with respect to each other under the influence of gravitational perturbations from other large bodies, especially the Sun and Jupiter.

[8] Horrocks's result for the Earth–Sun distance, published after his premature death in 1641 aged only 22, was 'at least 15,000 semidiameters of the Earth'. This corresponds to a lower limit of about 60 million miles, much larger than generally believed at the time, and highly controversial, but still about 30% smaller than the modern value.

[9] The mean Earth–Sun distance is known as the Astronomical Unit (AU), so called because the scale of the Solar System could be estimated in AU well before the distances between the planets were known in absolute terms.

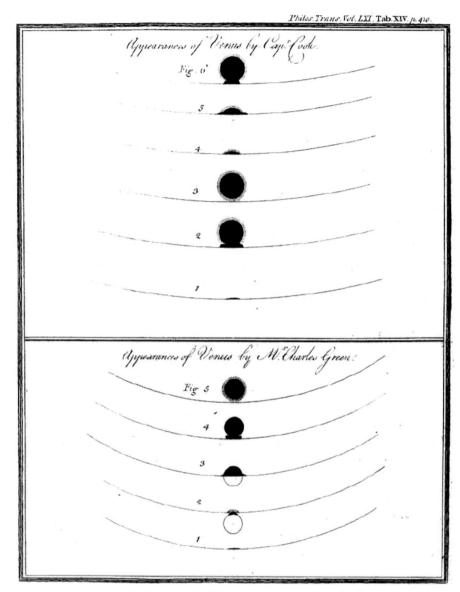

Figure 1.4 Captain Cook's drawing of the Venus transit of 1769.

from his base at Venus Point (still so called) in Tahiti, combined with others in Norway and Canada, yielded a value for the Earth-Sun distance of '93,726,900 English miles', which is correct to better than 1 per cent.

Early observations: another planet with an atmosphere

Dark, blotchy features on the disc of Venus had been reported by Cassini as early as 1666, and by other astronomers at various times since. When they thought they had seen something, it was generally assumed that these features were on the surface, or perhaps the combined effect of patches of cloud moving over the visible surface. Today, credit for discovering the atmosphere of Venus is generally given to a Russian observer, Mikhail Lomonosov, who recorded in the journal of the observatory at St Petersburg that the disc of Venus showed a halo during the solar transit of 1761. From this he deduced that the planet 'is surrounded by a considerable atmosphere, equal to, if not greater than, that which envelops our earthly sphere'.

At various times in the 1790s, the German astronomer Johann Schroeter reported observations of diffuse and variable markings on Venus, which he attributed to atmospheric phenomena. These markings may not have been real, but the limb darkening and the extension he saw of the 'horns' of the crescent Venus right around the planet probably were, and are also indications of a substantial atmosphere. Writing in 1793, William Herschel reported from his own observations of 'faint and changeable spots' that it was evident that Venus 'has an atmosphere'. These changes led him also to report that the fact that 'Venus has a motion on an axis cannot be doubted', although they 'surely cannot be on the solid body of the planet'. Indeed they are not.

Observations of surface features

As telescopes got better, and observers strained to see features on Venus that could tell them something about the nature of the nearest planet, reports of various phenomena filtered into the journals of professional scientific societies around the world. In the second half of the nineteenth century, these included occasional bright spots that were sometimes inferred to be snow-covered mountain peaks catching the sunlight. Bright polar caps were also seen by a large number of highly reputable astronomers using the latest instruments, and generally assumed to be icy like those on Earth and Mars. Today, transient bright clouds are seen on Venus from orbiting spacecraft and also from the Earth, and there is a spirited debate as to their cause, the two most popular theories being volcanic plumes or some as yet unexplained meteorological phenomenon. The polar caps are certainly present as well, but as semi-permanent features in the cloud cover, rather than ice on the surface. Whether the Victorian astronomers actually saw these through their telescopes is debatable; nowadays we require special photographic observing techniques not available before the 1920s.

In the summer of 1886, Percival Lowell, soon to become notorious for his interpretation of features seen on Mars as canals built by an intelligent civilisation, turned his new, state-of-the-art 24-inch refracting telescope on Venus. He had used his considerable wealth to build this facility on Mars Hill near Flagstaff, Arizona, in order to make better observations of the red planet and the civilised artefacts that excited him so much. However, when the new observatory was commissioned, Mars was not well located in the night sky over Arizona, so he looked at Venus instead. The markings he saw (Figure 1.5) seemed sufficiently reproducible for him to become convinced that he was observing features on the surface, providing 'evidence for slowness of rotation'. Unlike his vision for Mars, he did not claim the features looked artificial, describing them as 'perfectly natural' and the result of 'rock or sand weathered by aeons of exposure to wind and sun'.

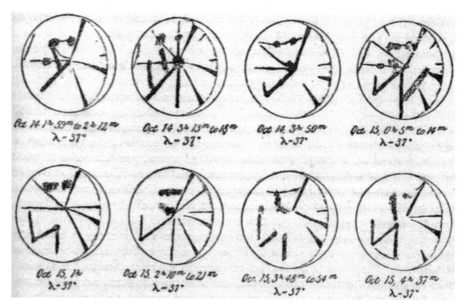

Figure 1.5 Sketches of Venus obtained by Percival Lowell from his observatory in Flagstaff, Arizona in August 1896, showing what he took to be surface features viewed through the 'brilliant straw-color veil' of Venus's atmosphere[10].

Other astronomers, including some of Lowell's own assistants using the same telescope, were sceptical, and reported that they could not see the features which Lowell described as 'perfectly distinct'. A plausible theory has been advanced which suggests that Lowell was observing the blood vessels in his own eye, reflected in the lenses of his telescope. He suffered, and in 1916 died suddenly, from high blood pressure, a condition that most likely made his retinal arteries more prominent than normal. This, plus his ambitious and excitable nature, may account for why Lowell thought he saw features on Venus that were invisible to others, and that we can now be certain do not exist.

The ashen light

One of the earliest discoveries about Venus was that the night side is not completely dark. Many observers, since the Jesuit priest Giovanni Riccioli of Bologna as long ago as 1643, have reported the emanation of a mysterious glow from the main disc when observing the planet at times when the sunlit side presents itself as a narrow crescent (Figure 1.6). The purported glow, which became known as the ashen light, is extremely faint and not always present even under good observing conditions. Some astronomers say they have

---

[10] This quote is from an article Lowell wrote in the German journal *Astronomische Nachrichten*, in 1897. There he went on to state emphatically that the markings on Venus were to him as distinct as those on the Moon, that they 'disclosed the rotation period unmistakably', and 'are not obscured at any time by clouds. In other words there are no clouds on the planet'.

Figure 1.6 A sketch of Venus by Patrick Moore, using a 15-inch reflector, showing the ashen light (his observing notes say that its brightness is exaggerated for clarity).

never been able to see it, and doubt its existence,[11] but they are far outnumbered by those who have. The latter describe it as being dark red or brownish in colour, patchy and changeable in shape and coverage, and variable in brightness.

There has been much discussion over the years as to what could be the source of this light, but there is still no general agreement. In the nineteenth century, there were imaginative suggestions involving celebratory bonfires and fireworks by hypothetical inhabitants of Venus. The 'leakage' of sunlight from the dayside, via scattering processes in the clouds, was a more realistic idea, but the bright regions are not concentrated near the terminator,[12] and it is hard to produce a model for the process responsible if this is the cause. More recently, it has become popular to posit that either some sort of auroral or airglow effect, or possibly frequent, widespread lightning discharges, must be responsible. However, close-up examination using television cameras on various spacecraft has failed to find the necessary evidence for this.

Some of the most recent investigations, from Earthbound telescopes and from instruments on spacecraft, have revealed initially surprising properties of the surface and the clouds that offer a new explanation for the mysterious emissions, probably the right one at last. We now know that the cloud layers on our sister planet, although extensive, are translucent at visible and near-infrared (IR) wavelengths, and it follows that we are seeing

---

[11] E. E. Barnard (1857–1923), who was professor of astronomy at the University of Chicago and who, among numerous other achievements, was the first since Galileo to discover a satellite of Jupiter (Amalthea), was one of those undoubtedly meticulous observers who were sceptical about the existence of the ashen light on Venus (and canals on Mars).

[12] The terminator is the day-night boundary line.

through them to the surface, which is glowing faintly with a dull red heat. The many reports of the colour of the night side of Venus as being a dim rusty shade lend further support to this idea. If it is correct, the observed patchiness and variability are both attributable to the large-scale structure of clouds in the deep atmosphere.

In order not to get too far ahead of the story, we will return to the topic of the ashen light and discuss this theory in more detail later, after covering more of the historical and scientific background. The implications turn out to be almost as exciting as observing Venusian firework displays from Earth.

The ultraviolet markings

Modern observing methods in the visible part of the spectrum show no markings on Venus that could be associated with surface features. Instead, with rare exceptions, most notably Lowell, a few visual observers have reported only subtle and ephemeral markings on the disc, and most see no contrast at all.[13] Some very faint streaks in the clouds, and slight 'scalloping' of the terminator line which separates the day and night sides, are the only irregularity commonly seen through the telescope with the naked eye (Plate 5).

Even the television cameras on the spacecraft *Mariner 10*, which observed Venus from a distance of 10,000 kilometres in 1973, were unable to detect significant contrasts over the brilliantly reflective disc of the planet when observing at visible wavelengths. Cameras, however, can image the planet at wavelengths to which the human eye is not sensitive, and they do not have to move very far into the ultraviolet (UV) spectrum, or in the other direction in wavelength to the near infrared, before the story changes completely.

The pictures of Venus that one usually sees, with prominent markings in the clouds, are taken photographically through a UV filter (Figure 1.7). At these short wavelengths,[14] the sulphur compounds in the clouds absorb sunlight, and differences in their density distribution within the upper cloud layers show up in the images as dark markings. Even these are quite subtle, and the contrasts are often stretched by a computer before the picture is released to the public.

The UV markings were first observed in the 1920s, but it was not until the mid-1980s, well into the era of space exploration, that it was discovered that striking contrasts can also be observed on the night side at certain wavelengths in the near-IR spectrum.[15] The UV observations are of course seen on the side of the planet that is illuminated by the Sun,

---

[13] However, Patrick Moore, probably the best and certainly the most famous amateur observer of recent times, states categorically (in 'Venus', London 2005) that 'it is wrong to say that nothing whatsoever can be seen on Venus using ordinary telescopes', and has reported his own observations of bright spots, polar caps and 'vague, elusive' shadings on many occasions.

[14] In planetary observations, wavelengths are usually measured in micrometres (one millionth of a metre), often abbreviated to microns or micron. Ultraviolet pictures of Venus are typically at about 0.35 micron, where visible light has wavelengths from about 0.4 to 0.8 micron.

[15] 'Infrared' means wavelengths longer than red light, i.e. lower in frequency than the lowest the eye can detect. The 'near' infrared is the part of this that is nearest to the visible in wavelength, typically from 0.8 micron, out to about 4.0 microns where the 'mid' infrared begins. The 'far' infrared corresponds to wavelengths from 20 microns, out to 1 millimetre where the microwave spectrum takes over. This

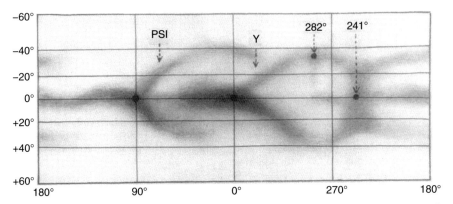

Figure 1.7 A sketch map of real quasi-permanent markings in the clouds on Venus, produced from ultraviolet photographs.

while the near-IR features are emissions from the planet itself and so dim that they can be seen only on the dark side. In this respect they are obviously related to the ashen light phenomenon, but were discovered at longer wavelengths, using IR detectors rather than the human eye. The cloud patterns that they reveal are different from those in UV images obtained at about the same time, even allowing for the time it takes for features to travel from the day to the night side (about two days). This suggests that the two types of observation are sensitive to different cloud layers at different heights in the atmosphere, which does seem to be the case, as we discuss later in Chapter 6. We will also see that we can, at last, get glimpses of the surface through the clouds by observing in the near infrared under certain conditions.

Speculation on the nature of the clouds and conditions at the surface

Long before Earth-based telescopes first revealed the ultraviolet markings in Venus's clouds, it had been realised that the brightness of Venus in the morning and evening skies, and the general lack of visible features, must be a result of a thick, uniform and permanent covering of clouds. The idea was consistent with the understandable supposition that Venus was much like the Earth in most ways, but with more evaporation from its presumed oceans because of the stronger solar heating, and therefore more cloud. Svante Arrhenius, in his book *The Destinies of the Stars* published in 1918, started with the fact that Venus's distance from the Sun is 72 per cent that of Earth. Using observational estimates of the reflectivity or albedo of the cloudy globe, he then worked out that the mean temperature at equatorial latitudes would be about 47°C, compared with 26°C for the tropics on Earth. Extrapolated to the poles, this sort of temperature would not be compatible with polar ice caps, which suggested to some astronomers that Venus could be entirely covered with oceans of water.

terminology is largely historic, having to do with the different technologies required to detect infrared radiation in these regions, but is still used extensively.

Others proposed that Venus might be dry and desertlike, in which case the clouds might be windblown dust or sand. The argument in this case was that, being close to the Sun, Venus had dried out over the aeons, like the once-fertile Sahara on Earth. If the clouds were dust, this could explain the slight lemon-yellow coloration that apparently distinguishes the Venusian clouds from their terrestrial counterparts.

However, until the space age began in the 1960s, the commonest picture of the surface of Venus in popular astronomy was one that resembled the primitive Earth, with some seas but also rain-soaked landmasses (Plate 6). It did not seem unreasonable to go on to invoke swamps, tropical vegetation, animal life and even humans. These might be cavemen fighting dinosaurs, or advanced civilisations living in cities and using technology. Both of these scenarios were popular in the fiction of the time, of which the best known today are the John Carter series of Edgar Rice Burroughs and the Pellucidar novels of C. S. Lewis.

The search for water vapour in the atmosphere

By the 1930s, it had become possible to make useful observations of the nearby planets using a spectrograph, an instrument that breaks light into its constituent wavelengths, like a rainbow but with finer detail and the ability to work at wavelengths beyond the range of sensitivity of the human eye. When attached to a good-sized telescope, these devices offered the possibility to detect the characteristic absorption features of common atmospheric gases, including water vapour. Percival Lowell and his team at Flagstaff had made strenuous attempts to show that the Martian atmosphere had a measurable humidity, as it would have to if there were open canals carrying liquid water across the surface. They came up with several false detections, but it was not until the 1950s that the Martian water vapour was definitely found, and at a much lower concentration than Lowell had hoped for.

If the clouds on Venus were similar to those on Earth and composed mostly of water droplets, then there ought to have been plenty of vapour for the early spectroscopists to detect there, too. In the United States, Walter Adams and Theodore Dunham were among the first to look, in the 1930s, and they did find interesting features in their spectra, although when they studied them in detail, the absorption lines turned out to belong to carbon dioxide.

Later observers, with gradually improved instrumentation and high-altitude observatories on aircraft, found many more bands of carbon dioxide lines at progressively longer wavelengths, and by the 1960s it was apparent that this gas was a major component of the overall atmospheric composition. At this time observers were also reporting water vapour detections, but these showed confusingly high levels of variability and were mostly at levels too small to be consistent with clouds made up of water droplets or ice, which was what most astronomers expected (Figure 1.8).

By 1972, Andrew Young and others had proposed that the spectrum of the clouds was best matched not by water but by sulphuric acid, and this was confirmed by the middle of the decade by the use of polarimetry, which allowed the refractive index of the cloud material to be inferred. It was consistent with a mixture of about three parts H_2SO_4 to only

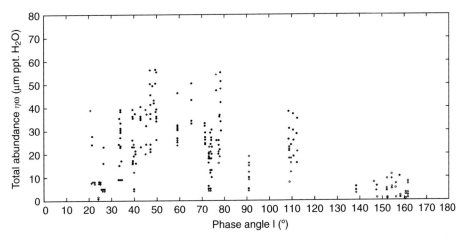

Figure 1.8 Earth-based observations of the column abundance of water vapour above the clouds as a function of phase angle (the angle between the Earth–Venus and Sun–Venus directions). The amounts are very small, and remarkably variable.

one part of H_2O – a very strong acid solution. Experience in the laboratory on Earth readily shows such a solution to be highly corrosive for many common materials that might be exposed to it, including, of course, human tissue. As if this were not enough by way of a slap in the face for the tropical paradise image of Venus, evidence was also building that the surface is almost unimaginably hot.

A hot and arid surface?

The first indications that the surface conditions on Venus might be inhospitable to human explorers came in the 1950s with the advent of the first large radio telescopes. In the passive mode normally used in astronomy, the large dishes tune in to microwave radiation – consisting of short radio waves with wavelengths typically measuring a few centimetres – emitted from the surface of the planet. Unlike the shorter-wavelength visible and infrared radiation, microwaves pass almost unaffected through the cloud layers, and the intensity of the emission from the surface can be measured on Earth. The intensity is related to the temperature of the emitting surface in a known way, through Planck's radiation formula. The early results for Venus, made in 1956 using a 50-foot dish at the US Naval Research Laboratory in Washington to observe at wavelengths in the range from 3 to 10 centimetres, corresponded to a source at a temperature of more than 300°C, much too hot for free water or plant life, let alone dinosaurs (or humans).

At first, such high temperatures for the surface of Venus did not seem reasonable; after all, the solar intensity is only twice that of the Earth, and most of that is reflected away by the bright clouds, back into space. In fact, because of the high albedo, the solar heating of Venus is actually *less* than Earth, and about the same as distant, chilly Mars.

Even if the atmosphere of Venus were much thicker than Earth's, it seemed at first incredible that it could trap enough heat to elevate the temperature that much through the greenhouse effect. Alternative explanations were sought, the most promising of which was that the microwave emission from Venus was not from a hot surface, but instead produced by non-thermal emission from electrons in Venus's ionosphere, approximately 100 kilometres above the surface.

Calculations soon showed that the ionosphere would have to be implausibly dense to match the observations, leaving the choice between two scenarios, both of which were seen as rather unlikely by most scientists. An exception was the young Carl Sagan, who was grappling with the problem as part of his doctoral studies at the University of Chicago. Sagan made calculations that convinced him, at least, that the greenhouse effect was responsible, and began to look for ways to gain experimental proof.

As the debate about Venus raged on in astronomical circles, the space age began with the launch of *Sputnik 1* on 4 October 1957. Both the Americans and the Soviets were keen to achieve successful flights to the nearby planets. The closest, Venus, was high on the list, along with Mars, rather farther and harder to reach but full of mystery and promise. Investigating the possible high temperatures on Venus made an obvious scientific goal for a mission, and would enhance the achievement of simply getting there.

There were basically two ways of doing this. The first was to fly a small microwave radiometer all the way to Venus, to investigate the emission phenomenon close up and verify whether the radiation being picked up from Earth really was from the surface. The other was to dive into the atmosphere, land on the surface and measure the temperature directly. Better still, do both. The race was on.

Chapter 2
Mariner and *Venera*: the first space missions to Venus

The Soviet Union launched the first space probe towards Venus on 4 February 1961. However, this failed, and so did their next several attempts. The Americans, too, came unstuck on their first attempt. It was not to be expected that such a sophisticated endeavour as the first flight to another planet would be achieved easily, and both teams soon tried again. In the end, it was the Americans who got a working spacecraft to Venus first.

The Venus *Mariners*: the first close-up views

The US space agency NASA was set up in 1958 and among its first tasks was the development of the *Pioneer* series of small spacecraft to explore the interplanetary medium near the Earth. These were followed by the *Surveyor* series, which targeted the Moon. A larger spacecraft than these would be needed to go on to even the closest planets, and NASA gave the job to its newly acquired centre in Southern California, the Jet Propulsion Laboratory (JPL). Before this, JPL had been an Army Air Corps facility for the development of rocket engines, with the name dating back to 1943. The new series of spacecraft was called *Mariner*, and Venus was its first target.

The scientific payload of *Mariners 1* and *2* consisted of microwave and infrared radiometers to observe the radiation emitted from the planet, and a magnetometer to detect its magnetic field, plus devices to measure charged particles, cosmic dust and solar plasma in the surrounding space environment. The star of the show, accounting for most of the mass, was the microwave radiometer, whose goal was to confirm or refute the high brightness temperatures of around 300°C that the Earth-based radio astronomers had measured at wavelengths in the microwave part of the electromagnetic spectrum.[1] It was also hoped that close-up measurements would reveal whether or not this high brightness was really due to an elevated surface temperature, or to non-thermal radiation from some kind of electrical discharge in the upper atmosphere, as had also been postulated.

---

[1] The 'brightness' temperature is that which corresponds to the intensity of the radiation emitted, using Planck's radiation law to connect the two. For a solid object (theoretically one that is perfectly black, but most real materials such as basalt rocks come close) this will approximately equal the actual temperature.

The approach to designing the instrument was to make measurements at several wavelengths, chosen for their ability to discriminate between models of Venus's atmosphere corresponding to different possible surface conditions. Microwave instruments tend to be heavy and it was possible to include only two wavelength channels in the end; even then, the radiometer weighed a hefty 21 pounds,[2] about the same as the other instruments combined.

The choice of the wavelengths to use required a great deal of thought since so little was known about Venus at the time (this was 1960). Carl Sagan and others had been devising model atmospheres to see if high surface temperatures could be explained by an extreme version of the familiar 'greenhouse' effect,[3] which traps the Sun's heat on the Earth and keeps the surface here around 35 degrees warmer than it would be otherwise. For Venus the warming effect would have to be more like 300 or 400 degrees, but Sagan showed this was possible under certain conditions, mainly if the planet has a thick atmosphere containing a lot of carbon dioxide. A range of these models with different values for the unknowns, such as surface pressure and water vapour abundance, was used to compute the expected microwave spectrum, and these were compared to the Earth-based measurements.

Figure 2.1 shows such a comparison: here, the composition has been fixed at 75 per cent carbon dioxide, 24 per cent nitrogen, and 1 per cent water vapour, representing the experimenters' best guess based on everything known at the time. The surface temperature was fixed at 300°C to match the brightness temperature at wavelengths where the absorption by the atmosphere was expected to be small, so the observed radiance should be nearly all coming from the surface. Finally, they guessed that the surface pressure lay somewhere between 2 and 20 atmospheres.[4] Looking at the plot, we see that two atmospheres does not really fit the data near 0.8 centimetres wavelength, and that something higher than 20 atmospheres would actually be better. At that time, even scientists had trouble imagining anything so different from the Earth, and they knew that the data and the models were error-prone.

Nevertheless, this study showed them how best to deploy the two available wavelengths. The sharp feature near 1.35 centimetres is the absorption due to a transition in the rotational energy of the water vapour molecule. This is hard to observe on Venus from the Earth because of all the water in our own atmosphere, but was obviously a good target for a mission to Venus. The figure of 1.9 centimetres was chosen as the other wavelength because most of the radiation there comes from the surface. In addition, it is reasonably close to 1.35 so that other wavelength-determined factors, such as the

---

[2] NASA used English units until the 1970s, and sometimes even after that. The loss of *Mars Climate Orbiter* in 1999 was famously due to a mix-up between English and metric units in calculating the trajectory.

[3] A 'model' atmosphere is an adopted, simplified representation of the temperature and pressure versus height, and the atmospheric composition, used in calculations to analyse data, predict behaviour, and plan further experiments. They can range from simple one-dimensional profiles to fully three-dimensional, time-dependent general circulation models.

[4] One atmosphere (abbreviated atm) is the mean surface pressure on Earth.

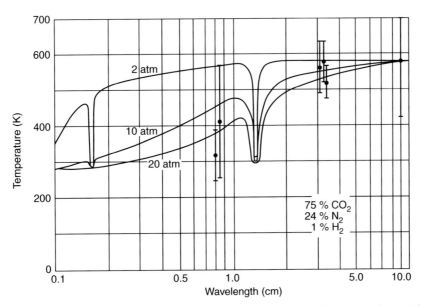

Figure 2.1 The calculated spectrum of Venus for three models of the atmosphere with surface pressures ranging from 2 to 20 atmospheres, compared with Earth-based measurements represented by the points with error bars.

diffraction-limited beam width of the radiometer,[5] would not complicate the comparison of the two readings.

Beam width was important for another reason. The *Mariner* scientists wanted to resolve the disc of Venus and scan across it. The way in which the brightness varied from the centre to the edge of the disc would be telling: if the radiation came from a hot surface, it would fall off towards the edge (limb darkening). If it was from the upper atmosphere, it would brighten towards the limb as the path length through the source layer increased. Achieving this crucial measurement meant keeping the wavelength as short as possible (and the instrument aperture as large as possible within the limited mass available, which translated to a dish about half a metre in diameter). It also meant, of course, that the spacecraft had to fly as close as possible to Venus without hitting it. This too was a challenge, with interplanetary navigation techniques still very much in their infancy.

Mariner 2

The first *Mariner* lifted off from Cape Canaveral on 22 July 1962, but the launch vehicle's guidance system failed and the rocket and its payload strayed off course. It had to be

---

[5] Instruments of this type accept radiation from the target that falls inside a roughly conical beam of a width that is determined by the design of the optics. The edges of this cone are fuzzed by interference between the incoming photons, a phenomenon known as diffraction.

deliberately destroyed by the controllers on the ground, only five minutes into what was meant to be a four-month flight to Venus. In those days spacecraft were built in pairs to allow for this kind of failure, so a little over a month later, on 27 August 1962, an identical spacecraft was launched successfully on an *Atlas* rocket with an *Agena* second stage (Figure 2.2).[6]

Mariner 2 became the world's first successful planetary mission when it flew past Venus at a distance of 35,000 kilometres on 14 December 1962. This approach was about 10,000 kilometres further away than the design goal, but less than the error the conservative engineers had predicted, so it represented a considerable achievement. Furthermore, the payload was working. The radiometer on board functioned well enough to confirm that the source of radiation at 1.9 centimetres is indeed very hot, even more than 300°C, and showed unambiguous limb darkening, meaning it was from the surface of Venus.

The other instruments also delivered intriguing findings. The infrared radiometer showed that there was little difference between the cloud-top temperatures on the dayside and nightside, which suggested to the investigators that there is a deep atmosphere stretching far below the clouds. The magnetometer could find no planetary magnetic field, certainly nothing as strong as Earth's, something of a surprise in view of the similar sizes and expected interior compositions of the twin planets.

Mariner 5

Mariners 3 and *4* went to Mars, so *Mariner 5* was the next American mission to Venus, launched on 14 June 1967 and flying past the planet on 19 October 1967, at a distance of less than 4,000 kilometres from the surface. Since it flew behind the disc, as seen from the Earth, the 'radio occultation' technique could be used for the first time, where the variation in the radio signal from the spacecraft is analysed as it passes through the atmosphere. The line of sight between the communications antenna on the spacecraft and the receiver on Earth bends due to refraction as it passes through the atmosphere from top to bottom. It also changes in frequency, due to the speed of light in gas being different from that in the vacuum of space. Using some basic laws of physics these signal changes can be interpreted in terms of the temperature and pressure of the atmosphere as a function of height.

The results for the rate at which the pressure fell with altitude in the upper atmosphere provided a new estimate of the mean molecular weight of the gas mixture. Since this was higher than expected from the pre-*Mariner* model value of around seventy-five, the concentration of the heaviest abundant gas, carbon dioxide, was revised up to 90 per cent.[7] The highest temperature measured was about 180°C, but since the occultation method

---

[6] *Atlas* was originally developed by the Convair Corporation as an intercontinental ballistic missile to deliver nuclear warheads, before it became a NASA workhorse for launching satellites and planetary probes. A version of it (*Atlas V*) is still in use. *Agena* was developed by Lockheed as a small satellite launcher and later as an upper stage for larger rockets like *Atlas*.

[7] The actual value is around 95 per cent.

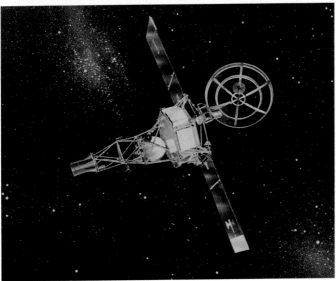

Figure 2.2 The *Mariner 2* spacecraft and its *Atlas-Agena* launch vehicle.

works only down to the middle of the atmosphere, the exact height depending on the surface pressure, this was again consistent with a surface temperature of at least 300 degrees. The signal actually ceased at a height of 32 kilometres, when the transmission path was bent by 17 degrees by the atmosphere but almost totally absorbed. Extrapolating the values downwards from there led to values for the surface temperature in the range 375 to 800°C, and a surface pressure somewhere between 60 and 100 atmospheres, when the various errors and uncertainties were taken into account.

Mariner 10

Mariners 6 and *7* were Mars flyby missions in 1969, and *Mariner 9* became the first to orbit the red planet in 1971 (*Mariner 8* having failed at launch). Thus, it was not until *Mariner 10* that Venus was in the spotlight again. This mission, known before launch as *Mariner-Venus-Mercury*, performed a flyby of Venus on 5 February 1974, using a 'gravity assist' from this manoeuvre to fly on to Mercury, making close encounters with the innermost planet three times between 1974 and 1975 as it circled the Sun.

Mariner 10 carried a sophisticated camera, so the Venus encounter offered the chance to carry out the first detailed photographic survey of the planet from close range. The experimenters knew from the Earth-based experience that they were unlikely to see very much at visible wavelengths, so the camera was equipped with filters at six different wavelengths – orange, blue, ultraviolet, ultraviolet polarising, blocking and clear. As anticipated, the ultraviolet filter delivered the best contrast, and revealed details in the cloud structure on all scales from planetwide to the smallest details observable, a few kilometres across (Plate 7).

The big 'sideways Y' feature that Earthbound observers had followed for decades was there (Figure 2.3), but so also was a host of more intricate structure, including cloud

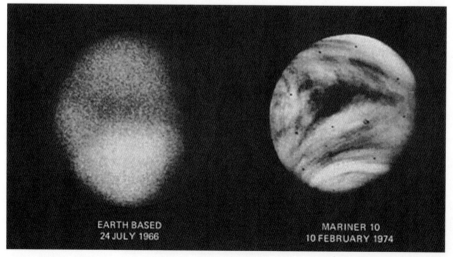

EARTH BASED
24 JULY 1966

MARINER 10
10 FEBRUARY 1974

Figure 2.3 The features seen in Venus's cloud tops through an ultraviolet filter from the Earth and by the camera on *Mariner 10*. (Compare this with the sketch in Figure 1.7.)

patterns suggesting waves and turbulence, and some rather strange linear features. The 'Y' was revealed as a combination of large-scale planetary waves that seem to embrace the subsolar point, where convective, 'boiling' activity is seen to be present in the cloud patterns. The rapid circulation of the ultraviolet-dark features around the equator at speeds of over 100 metres per second was also quite obvious in sequences of images taken several minutes apart.

The cloud-tracking experiment also showed meridional (equator-to-pole) motions, much slower than the zonal ones (east to west), and dark features spiralling in towards the poles themselves, where the rapidly circulating air descends in a giant vortex (Figure 2.4). All these features and more, just glimpsed by *Mariner 10* during its rapid flyby, marked the beginning of Venus meteorology as a serious science. Even better, they showed how we could investigate the atmospheric motions and the cloud features in more depth in future once we had cameras and other instruments on an

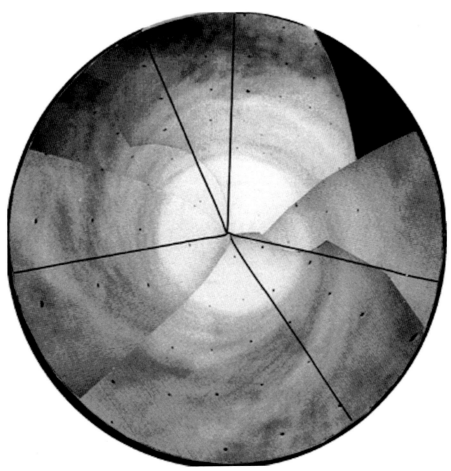

Figure 2.4 A mosaic of ultraviolet images from *Mariner 10*, centred on the south pole and showing a suggestion of large-scale vortex activity.

orbiting spacecraft that could take continuous data for months or years rather than a few hours.

Venera: the first landings

In the politically fraught days of the Soviet Union, the Russians were even more reticent about discussing their achievements and plans, and especially their failures, than they are now, so many of the details about the early days of their Venus exploration programme have become public only fairly recently.[8] Communications between scientists working on either side of the Iron Curtain were, in those days, mainly through meeting at international conferences, especially the regular General Assembly of the Committee on Space Research. Known to all as COSPAR, this fine organisation, working from its headquarters in Paris, has always sought to promote international collaboration on all aspects of space science and technology, and the related politics. Its biannual General Assemblies move around the globe between continents and political blocs, most recently from Warsaw to Houston, Texas; from Paris to Beijing, from Bremen in Germany to Mysore in India. It was exciting to find that some of the Russian scientists were friendly to their western counter-parts, and would chat discreetly about what they were doing when the grim-faced men in raincoats were not around .[9]

The first attempt to land instruments on Venus was as early as 4 February 1961, but this and a number of successors all failed, either due to launch problems or from loss of communication at some point during the long flight. It was not until 1 March 1966 that *Venera 3* became the first man-made object to impact the surface of another planet, although its scientific payload malfunctioned so it returned no data. *Venera 4* fared better, and succeeded on 16 October 1967 in delivering the first results on temperature and pressure in the atmosphere. The sensors were reading a temperature of 275°C at a pressure of about 20 atmospheres when the signal stopped abruptly; it was assumed that *Venera 4* had hit the surface. The *Mariner 5* investigators later pointed out that in their radio occultation profile, these values would lie at a height of about 25 kilometres above the surface, assuming the latest value for the radius of Venus obtained from radar data. Extrapolating the profile down to the surface gives a temperature there that is similar to the *Mariner* estimate and close to the modern value.

The first scientifically successful landings were by *Veneras 5* and *6* (Figure 2.5), on 16 and 17 May 1969. Once down, they both sent data, including atmospheric composition,

---

[8] Even now, the information on official space history sites is brief and often speculative. See for instance that sponsored by NASA called 'Tentatively Identified Missions and Launch Failures', http://nssdc.gsfc.nasa.gov/planetary/tent_launch.html

[9] One of them was Vasili Moroz, who was the founder and leader for 37 years of the Department of Planetary Science at the Institute for Space Research in Moscow, an outstanding scientist and a great human being, who died in 2004.

Figure 2.5 The *Veneras 5* and *6* spacecraft were identical, weighing just over a ton at launch. The surface modules, both of which landed on Venus in May 1969, accounted for about a third of the launch mass, the rest being the carrier spacecraft with its power and communications systems.

temperature and pressure, from the surface, each surviving for just under an hour before being overcome by the high temperatures and pressures. Now there could be no possible doubt that the surface was very hot, and there was a measured value for the surface pressure at last – nearly 100 Earth atmospheres. The two go together – the high pressure is the main reason for the hellish temperature, and the reason the environment differs so much from Earth.

The later Soviet missions to the surface, starting with *Venera 9* on 22 October 1975, obtained remarkable photographs of the terrain (Figure 2.6). These pictures were obtained in natural light, when it was found that enough of the sunlight incident upon Venus could diffuse through the clouds to the surface to illuminate the

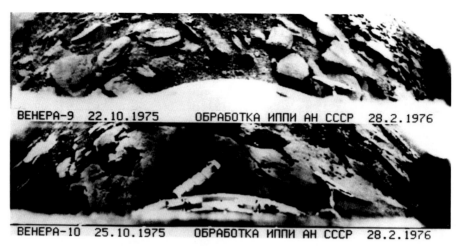

ВЕНЕРА-9 22.10.1975   ОБРАБОТКА ИППИ АН СССР   28.2.1976

ВЕНЕРА-10 25.10.1975   ОБРАБОТКА ИППИ АН СССР   28.2.1976

Figure 2.6 The first surface panoramas from *Veneras 9* and *10* showed plains covered with soil and rock at both sites. The boulders, probably derived from solidified lava flows, show various degrees of weathering.

scene.[10] In 1978, radiometers on the *Pioneer Venus* probes measured the solar flux and found that 2 per cent of the total falling on the planet actually reaches the surface without being absorbed.

This is much less than on Earth, where the figure is about 50 per cent, but surprisingly large considering the thickness and cloudiness of the atmosphere. In fact, it could have been realised from the brightness of Venus as seen from outside that the cloud droplets must be very reflective on the whole. They diffuse the radiation thoroughly by scattering each photon dozens of times during its passage through the atmosphere, but do not absorb as strongly as terrestrial clouds would, because they have a different composition and hence different optical properties.

The first photographs had been eagerly awaited, and were quickly released to the world as evidence of the 'splendid achievement of Soviet astronautics' that was the *Venera* programme. Since evidence about the high temperature had accumulated, the dreams of those who expected oceans and forests had evaporated, and the surface of Venus was duly revealed as a sterile, scorched desert. The *Venera 9* landing site is dominated by the boulders that are seen strewn about the landscape, but 2,000 kilometres away the *Venera 10* panorama shows a flat, rocky plain with a low outcropping of rock. It seems likely that *Venera 9* landed on a slope (sensors on the spacecraft showed that it was tilted at about 30 degrees to the horizontal after settling on the surface) and that the boulders are 'scree' or rubble from the break-up of the face

---

[10] The Russian scientists told us at the time that they had floodlights on the spacecraft, because they expected darkness, but in the event they weren't used. They compared the natural illumination at the landing site on Venus, with the Sun 60 degrees above the horizon, to 'Moscow during a thunderstorm'.

of the hill on which the spacecraft sits. *Venera 10,* on the other hand, sits on a fairly flat, rocky plain.

Both areas contain clear evidence of past volcanic activity. The cracked plains around *Venera 10* cannot be anything but layers of solidified lava. The rounded stones lying on the ground near *Venera 9* look a lot like the boulders in a rushing mountain torrent on Earth, but are more likely to be fragments of frozen lava that have been eroded by some kind of weathering process. Some of the rocks have dark bands running through them, like the famous 'hamburger' to the left of centre in the upper picture of Figure 2.6, and others have a patchy appearance, both signs of a mixed composition. The mottled deposit seen in places looks like soil made of crushed basaltic (volcanic) rock. The blackest regions in the background are so dark they are unlikely to be simple shadows or dark minerals but more likely depressions or fissures between the rocks, due to fracturing by some geological process. The sharp edges on some rocks confirm that they are broken fragments.

On Earth, such evidence of fracturing would suggest that the processes responsible were recent, because erosion by running water, large daily or seasonal temperature changes and physical erosion by wind-blown dust, would together soon soften and eliminate the evidence. On Venus, however, these forces might be less extreme and it is possible that the fracturing could have happened long ago. On the other hand, the rounded shape of some of the rocks is evidence of erosion, and forces not usually found on Earth might act powerfully on Venus. These could include chemical attack by acidic vapours in the atmosphere and melting of volatile components of the rocks, although exactly (or even, to be honest, approximately) what is going on remains one of the major unsolved questions about Venus to this day.

In March 1981, *Veneras 13* and *14* touched down in Phoebe Regio (Figure 2.7),[11] carrying the first working colour cameras.[12] Both landed on flat plains of solidified volcanic lava, but the *Venera 13* site is thought to be older, since it has apparently been buckled into shallow ridges by movements of the crust to produce a suggestion of low hills in the distance (Figure 2.8). Like the earlier scenes, the closest terrain is fractured and eroded to form dark gravel patches. The plain where *Venera 14* landed is flatter and has no fine material deposits, suggesting it is geologically younger. This means it was probably covered by fresh lava only a few million years ago, compared with hundreds of millions for the older site.

The rocky surface under the landers was studied with a number of ingenious devices, some devised to survive the heat so they could be mounted on the outside of the space-craft. Special alloys and lubricants had to be developed, and joints and bearings designed to give the correct clearances when the metal parts expanded under the hot conditions on Venus. The engineers building and testing the equipment had a problem because it would

---

[11] Phoebe is one of the female Titans in Greek mythology.

[12] *Veneras 11* and *12* had colour cameras, but produced no pictures for the frustrating reason that the protective lens caps failed to come off after landing. They were meant to be blown off by small pyrotechnic devices, but the pressure failed to equalise before these were ignited and the high outside pressure held the caps on.

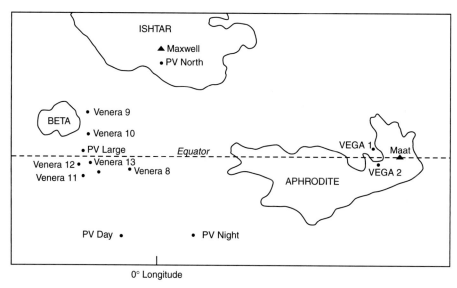

Figure 2.7 The approximate locations of the *Veneras 9* to *13* landing sites, and those for the later *VEGA 1* and *2* and *Pioneer Venus* (PV) Large, Day, Night and North probes discussed in Chapter 3, are shown relative to the three largest continents and the two highest mountains on Venus.

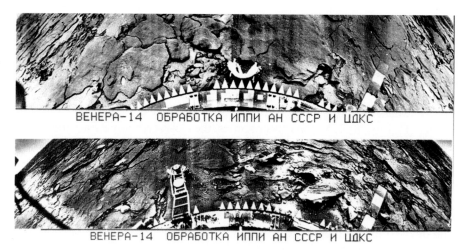

Figure 2.8 *Venera 14* images of volcanic plains in Phoebe Regio. Obtained in 1981, these are still the best close-up pictures of the surface of Venus that we have.

not work properly under normal lab conditions; all of the testing had to be done in massive ovens that were able to get much hotter than those in an ordinary kitchen.

Thus equipped, the probes hammered and drilled the adjacent rocky surface. It turned out to be less hard than it looks in the photographs: it is actually layered and crunchy.

Some of the geologists described it as resembling pumice, suggesting it was laid down in a cascade of volcanic ash rather than as a liquid lava flow.

The drills collected material from a few centimetres down and transferred it into a sample chamber. The transfer was achieved using the high pressure outside to force the sample from the drill core into the analyser, an X-ray fluorescence spectrometer excited by radioactive isotopes of plutonium and iron. To work properly, the sample chamber needed to be at a low pressure, which meant using more ingenuity to evacuate most of the atmospheric gas that it now contained. The solution was to seal the chamber and then open a valve connecting it to a large bottle containing a vacuum. This dropped the pressure over the sample from nearly 100 atmospheres to just 2 millibars.

The results at both sites indicated a composition similar to terrestrial basalts, confirming that it is material from volcanoes. Basalt comes in many variants on Earth, and it is not the same everywhere on Venus. For example, a higher content of potassium was found in the basaltic rock at the *Venera 14* site than at the others. Other instruments on the later *Venera* landers, principally gamma-ray spectrometers, provided information on the density and chemical composition of the surface rocks where they landed, finding that the abundances of the naturally radioactive elements uranium, thorium and potassium were again consistent with a composition within the range found in volcanic rocks on Earth.

The estimates of the density of the rock, by gamma-ray backscatter measurements, fell in the range expected for basaltic material. Basalt is the commonest type of rock in the Earth's crust, because it has a relatively low melting point so that it migrated to the outermost layer when the planet was condensing from a molten protoplanet into shells. This is also why it forms much of the lava that is extruded from volcanoes, not just on Earth but apparently also on the Moon, Mars and Venus.

Venera atmospheric measurements

All of the Soviet landers made atmospheric measurements during their descent to the surface. The instruments they carried included mass spectrometers and gas chromatographs to analyse the composition, and spectrometers and photometers to find out how the cloud layers affected the sunlight penetrating into Venus's atmosphere. Information on the vertical profile of these quantities, including the concentration of water vapour, could be gathered by repeated measurements as the probes descended. They confirmed that present-day Venus is very dry, with a measured humidity of just 30 parts per million in the lower troposphere. There was a suggestion of much larger and more ambiguous values at higher altitudes near the cloud decks, but these are less certain because the very presence of the clouds makes the measurements even more difficult and error-prone.

The first Venus orbiters

Placing a spacecraft in orbit around a distant planet is much more difficult than flying past. The navigation has to be a lot better, with the trajectory at closest approach finding the narrow window between crashing in to the planet and missing altogether. Clearly, a flyby is much more forgiving in this regard. In addition, it is necessary to have some means of slowing the spacecraft down at the right moment during the encounter, otherwise the window for attaining orbit does not exist at all. The rocket that provides the necessary impetus, and its fuel, has to be launched from Earth and carried all the way to Venus for a few minutes' use. Finally, the orbiter generally needs more sophisticated power, stabilisation, and communications systems than a spacecraft designed just to fly past the target. All of these refinements add more mass, leaving less for the scientific instruments that are the primary reason for going.

The Soviets, with their huge *Proton* launchers,[13] were the first to attempt a satellite of Venus, and in October 1975 they had success with two, *Veneras 9* and *10*. Before entering orbit they released the landers, which, as described previously, were the first to photograph the surface of Venus. They carried their own payload of scientific instruments, mostly optical devices such as cameras, spectrometers and radiometers, for studying the atmosphere from orbit, but they also served as relay stations for the probes, whose transmitters were incapable for reasons of power and line-of-sight geometry of reaching Earth directly.

Veneras 11 and *12* (another identical pair), which flew in 1978, reverted to flyby platforms for the carrier spacecraft that delivered the landers. Part of the reason for this was that launch opportunities to Venus, which generally occur every 18 months or so when the planets are suitably aligned, are not all the same. The 1978 missions arrived at Venus with three times the kinetic energy that their predecessors had, and would have needed a correspondingly larger orbit insertion motor with more fuel. By careful timing, even without sticking around in orbit, they were still able to act as relays for the landers before vanishing over the horizon.

Veneras 13 and *14* in 1981 were also flyby/landers. This time the flight dynamics were more favourable, but the payload on the surface had been increased by 100 kilograms, to allow more sophisticated experiments such as colour photography and subsurface drilling, and this made orbiting impractical.

It was not until *Veneras 15* and *16*, in 1983, that we saw another pair of orbiters from the Soviets. In a reversal of emphasis between the two main components of the mission, this time the planners had allowed for a massive orbiter payload and no lander at all. The reasoning here was to make it possible for the orbiter to carry an imaging 'synthetic aperture' radar, capable of penetrating the clouds and mapping the surface at high resolution (around one kilometre).

As well as the device itself, this would require an antenna in the form of a parabolic dish a metre and a half across (Figure 2.9). With this, the other instruments and more

---

[13] The Americans, of course, had the (even larger) *Saturn V*, but these were all reserved for the *Apollo* Moon landing programme that was going on at this time. Although there were some detailed studies of their subsequent use for launching missions to the planets, this never happened and the last launch by *Saturn V* was to carry the manned *Skylab* into Earth orbit on 14 May 1973.

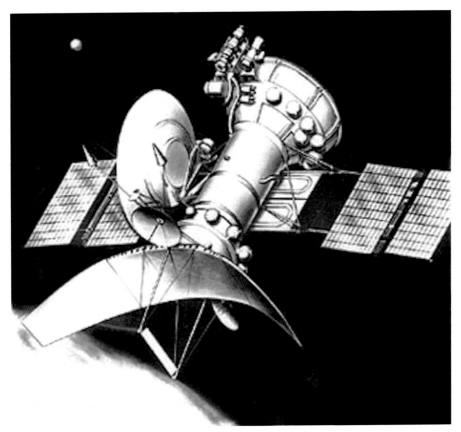

Figure 2.9 Drawing of the *Venera 16* orbiter in flight. The curved plate at the bottom is the antenna for the synthetic aperture radar, the dish at centre is for communicating with Earth.

than a ton of propellant required for orbit insertion, the spacecraft weighed in at a mighty four tons, and there was no mass left for a lander on this occasion. The radar experiment and the spectacular images of the surface that this mission obtained will be discussed in Chapter 4.

Chapter 3
Pioneer Venus and *Vega*: orbiters, balloons and multiprobes

The main achievement of the *Mariner* flybys and the *Venera* landings that took place between 1963 and 1983 was to show us the general nature of the Venus atmosphere and surface for the first time. Like all good experiments, they posed almost as many new questions as they answered. They therefore stoked controversy when it came to the interpretation of what clearly are very complicated, time-dependent phenomena, such as global circulation and weather in the atmosphere, and erosion and volcanism on the surface.

Spacecraft that fly quickly past the planet, and probes that descend rapidly through the atmosphere to die soon after on the surface, are the simplest missions to implement and are great for an initial exploration. However, to really understand what is going on, on a complex world, better coverage in space and time is needed than is obtained by a few scans across the planet or one vertical profile in the atmosphere. The second phase of Venus exploration would therefore require coordinated measurements from planet-circling orbiters for mapping and monitoring of the global atmosphere and surface, and of the near-space environment filled with neutral and charged particles and magnetic and electric fields. Ideally, these would be complemented by large numbers of simultaneous probes at different longitudes and times of day and night.

It would be nice to deploy long-duration stations and even rovers on the surface, as has already been done at Mars, but the problem of surviving the extreme Venusian surface environment for more than about an hour has not yet been solved. These ambitious missions are better thought of as part of a third phase still in the dim and distant future. An attractive and more feasible alternative would be balloons or aircraft floating in the atmosphere above and around the mountains and valleys. However, these are currently constrained to quite high altitudes, not lower than about 50 kilometres, in order to find conditions sufficiently benign to survive. But they could drop a series of small probes from there to sound the atmosphere below, and perhaps photograph or even dig in to and sample the surface itself. This has been seriously considered but so far not implemented. But again we are getting ahead of the story.

NASA plans multiple-entry probes and an orbiter

After *Mariner 10*, which was in fact primarily a mission to Mercury,[1] NASA had a lull on Venus while it focused on the ambitious *Viking* programme that was to land on Mars in 1976. Pressure from scientists interested in a comprehensive programme of Venus exploration, which became seriously organised in 1967, led to a string of NASA-supported studies of various options, some of them in collaboration with the fledgling European space organisation, then called ELDO.[2]

At that time, the *Mariners* and *Veneras* had shown that Venus had a dense, dynamic atmosphere primarily composed of carbon dioxide, a seriously un-Earthlike surface climate, and no magnetic field, but precious little else was known. NASA established a Science Steering Group in January 1972, charged with defining the best approach to addressing the major mysteries posed by Venus. Its report listed 24 questions that it would like to see answered by a suitable mission, to be carried out in the next few years. The list is worth reproducing here (Figure 3.1), as it forms a very concise summary of the scientific questions about the planet that were at the forefront at the time. Indeed, although *Pioneer Venus*, as the next mission was eventually named, would make progress on each of the topics (except perhaps No. 13, a curious inclusion), the list remains a reasonable blueprint for Venus exploration programmes to this day.

Pioneer Venus

In 1974 NASA signed off on a new start for what was to be its most ambitious Venus mission yet, originally targeting the launch opportunity in 1976, but eventually slipping to the next launch window in late 1978. The agency assigned the project to its Ames Research Center near San Francisco in Mountain View, California, which worked with the scientists to devise the mission design that would provide the best answers within the imposed cost ceiling of $200 million. The problem they faced was how to reconcile the different requirements to meet such a diverse set of goals.

Many of the objectives required an orbiter, to map the planet in the ultraviolet, visible and infrared, to make radar observations of the surface, and to explore the magnetosphere. On the other hand, things such as the cloud layering and particle properties were best measured from entry probes. Once it was agreed to have probes, there was a trade-off to be made between multiple small probes that gave better coverage, or one Soviet-style large probe with more, and larger, scientific instruments in the payload.

---

[1] *Mariner 10* needed the gravitational assist from a Venus flyby to get onto an efficient trajectory to Mercury, so the data from Venus were a bonus rather than an objective for the mission.

[2] The European Launcher Development Organization worked from 1964 to 1972 to develop a launch vehicle based on the British *Blue Streak* missile as a first stage, with French, German and Italian components on top of that. After some early successes the programme ran into trouble, despite a strong performance from *Blue Streak*, leading among other disappointments to withdrawal from the cooperation with NASA on plans for a US–European Venus mission.

1. Cloud layers: What is their number, and where are they located? Do they vary over the planet?
2. Cloud forms: Are they layered, turbulent, or merely hazes?
3. Cloud physics: Are the clouds opaque? What are the sizes of the cloud particles? How many particles are there per cubic centimeter?
4. Cloud composition: What is the chemical composition of the clouds? Is it different in the different layers?
5. Solar heating: Where is the solar radiation deposited within the atmosphere?
6. Deep circulation: What is the nature of the wind in the lower regions of the atmosphere? Is there any measurable wind close to the surface?
7. Deep driving forces: What are the horizontal differences in temperature in the deep atmosphere?
8. Driving force for the 4-day circulation: What are the horizontal temperature differences at the top layer of clouds that could cause the high winds there?
9. Loss of water: Has water been lost from Venus? If so, how?
10. Carbon dioxide stability: Why is molecular carbon dioxide stable in the upper atmosphere?
11. Surface composition: What is the composition of the crustal rocks of Venus?
12. Seismic activity: What is its level?
13. Earth tides: Do tidal effects from Earth exist at Venus, and if so, how strong are they?
14. Gravitational moments: What is the figure of the planet? What are the higher gravitational moments?
15. Extent of the 4-day circulation: How does this circulation vary with latitude on Venus and depth in the atmosphere?
16. Vertical temperature structure: Is there an isothermal region? Are there other departures from adiabaticity? What is the structure near the cloud tops?
17. Ionospheric motions: Are these motions sufficient to transport ionisation from the day to the night hemisphere?
18. Turbulence: How much turbulence is there in the deep atmosphere of the planet?
19. Ion chemistry: What is the chemistry of the ionosphere?
20. Exospheric temperature: What is the temperature and does it vary over the planet?
21. Topography: What features exist on the surface of the planet? How do they relate to thermal maps?
22. Magnetic moment: Does the planet have any internal magnetism?
23. Bulk atmospheric composition: What are the major gases in the Venus atmosphere? How do they vary at different altitudes?
24. Anemopause:[3] How does the solar wind interact with the planet?

Figure 3.1 Top science questions for Venus, according to NASA in 1972.

The final plan called for two launches on separate *Atlas-Centaur* vehicles in the same window of opportunity. The first would be the orbiter, to study the cloud-top morphology all over the planet and try to probe the cause and extent of the four-day 'super-rotation' at the cloud tops. A range of instruments would measure the vertical temperature structure and composition, including those in the ionosphere, the low-density region at high altitudes where normally stable molecules like water vapour and carbon dioxide are dissociated by radiation from the Sun. The orbiter would also carry a radar altimeter to map the surface topography by transmitting pulses of microwaves through the clouds and detecting the small fraction that returns.

The second launch would be a 'multiprobe' bus carrying four instrumented probes, three identical small ones and one more comprehensively equipped large one. Together they would descend through the atmosphere, slowed first by friction on their heat shields, and then by parachute. This 'one large plus three small' set of probes addressed the compromise between having the greatest number of probes spread across the planet

---

[3] This term was coined in the 1960s to describe the point at which the solar wind is blocked by the planet, either as a result of deflection by the planetary magnetic field or interaction with the upper atmosphere if there is no field.

to cover low and high latitudes, day and night, and having a substantial payload of instruments on each. The large probe carried the heavier instruments, such as the mass spectrometer that would measure atmospheric composition, while the small probes were restricted to smaller, simpler devices for measuring temperature, pressure and light levels.

The carrier bus itself would also have to collide with Venus, since it lacked any means to steer away after releasing the probes, but it had no heat shield or parachute to allow a controlled descent. The mass budget allowed for a couple of instruments – a neutral mass spectrometer and an ion mass spectrometer – to be mounted on the bus to make additional upper atmosphere measurements before burning up.

The probes were designed to investigate the detailed vertical structure of the clouds on Venus, their layering, microstructure and composition. Upwards- and downwards-viewing radiometers could measure the solar heating of the atmosphere as a function of depth, and tracking the drift of the probes would obtain wind profiles, to help work out the atmospheric circulation and its driving forces. The large probe would make new and hopefully better measurements of the bulk composition of the atmosphere, focusing on trying to understand the loss of water and the stability of carbon dioxide (the fact that solar radiation converts CO_2 to CO in the upper atmosphere at a great rate had raised the question of why the atmosphere is not by now mostly composed of the latter). Also targeted were the deep atmosphere, especially the vertical temperature structure near the surface, and the upper reaches where ionospheric turbulence, ion chemistry and exospheric temperature were goals to be investigated.

The total complement of scientific instruments on all six spacecraft is summarised in Figure 3.2. The instruments are grouped according to their principal scientific objective, and the code names begin with O, L, S or B to signify mounting on the orbiter, large probe, all of the three small probes, or probe carrier bus, respectively.

Had NASA been able to stick to the original schedule, they would have achieved the first American artificial satellite of Venus at the same time as the Soviets did with *Veneras 9* and *10*. Once the project got under way, however, schedule pressures caused the NASA launches to slip to the next window of opportunity, and it was not until 20 May 1978 that *Pioneer 12* (as it was officially designated) set off on its voyage. *Pioneer 13*, the bus carrying the four entry probes that would be the first US spacecraft to land on Venus, followed it on an identical *Atlas-Centaur* launched from Cape Canaveral on 6 August 1978.

Both spacecraft used the same basic body, a flat cylinder 2.5 metres in diameter and just over 1 metre high, built by the Hughes Aircraft Corporation in Long Beach, California (Figure 3.3). The orbiter was stabilised by spinning once every 12 seconds, which had the extra advantage of providing a scanning platform for those experiments, such as the particles and fields instruments, whose measurements required pointing that covered all directions. The orbiter weighed just over half a ton when launched, including 55 kilograms of scientific instruments and 200 kilograms of propellant for the orbit insertion motor. The multiprobe spacecraft weighed 875 kilograms, most of which was accounted for by the probes themselves at 585 kilograms (Figure 3.4).

Spin-stabilisation did make it more difficult for the 1-metre diameter communications dish to keep pointing towards the receiving station on the ground, however. This problem

Composition and Structure

LNMS	Mass spectrometer	J. Hoffman/University of Texas at Dallas
LGC	Gas chromatograph	V. Oyama/Ames Research Center
BNMS	Neutral mass spectrometer	U. von Zahn/University of Bonn
ONMS	Neutral mass spectrometer	H. Niemann/ Goddard Space Flight Center
OUVS	Ultraviolet spectrometer	I. Stewart/University of Colorado
L/SAS	Atmospheric structure	A. Seiff/Ames Research Center
OGPE	Atmospheric propagation	T. Croft/SRI International
OAD	Atmospheric drag	G. Keating/Langley Research Center

Clouds

LN/SN	Nephelometer	B. Ragent/Ames Research Center
LCPS	Cloud particle size spectrometer	R. Knollenberg/Particle Measuring Systems
OCPP	Cloud photopolarimeter	L. Travis/Goddard Institute for Space Studies

Thermal Balance

LSFR	Solar flux radiometer	M. Tomasko/University of Arizona
LIR	Infrared radiometer	R. Boese/Ames Research Center
SNFR	Net flux radiometer	V. Suomi/University of Wisconsin
OIR	Infrared radiometer	F. Taylor/Jet Propulsion Laboratory

Dynamics

DLBI	Long baseline interferometry	C. Counselman/MIT
MWIN	Doppler tracking of probes	A. Kliore/Jet Propulsion Laboratory
M/OTR	Atmospheric turbulence	R. Woo/Jet Propulsion Laboratory

Solar Wind/Ionosphere

BIMS	Ion mass spectrometer	H. Taylor/Goddard Space Flight Center
OIMS	Ion mass spectrometer	H. Taylor/Goddard Space Flight Center
OETP	Electron temperature probe	L. Brace/Goddard Space Flight Center
ORPA	Retarding potential analyzer	W. Knudsen/Lockheed Palo Alto Research
OMAG	Magnetometer	C. Russell/University of California at Los Angeles
OPA	Plasma analyzer	J. Wolfe/Ames Research Center
OEFD	Electric field detector	F. Scarf/TRW
ORO	Dual-frequency occultation	A. Kliore/Jet Propulsion Laboratory

Surface and Interior

ORAD	Radar mapper	G. Pettengill/MIT
OIDD	Internal density distribution	R. Phillips/Jet Propulsion Laboratory
OCM	Celestial mechanics	I. Shapiro/MIT

High-Energy Astronomy

OGBD	Gamma burst detector	W. Evans/Los Alamos Scientific Laboratory

Interdisciplinary Scientists

	Atmosphere and Geophysics	G. Schubert /University of California
	Fields and Particles	A. Nagy/ University of Michigan
	Geology and Geophysics	G. McGill/ University of Massachusetts
	Surface Features	H. Masursky/US Geological Survey
	Atmosphere	T. Donahue/ University of Michigan
	Atmosphere	J. Pollack/NASA Ames Research Center

Figure 3.2 Scientific experiments and principal investigators on *Pioneer Venus*.

was solved by 'despinning' the antenna, which meant mounting it on a ball race bearing and driving it to rotate in the opposite direction to the spacecraft, so it stayed on line with the Earth. This was something that had been done before and is not as difficult as it might sound. Anyway, it worked perfectly.

The orbiter arrived at Venus on 4 December 1978, followed by the probes, only four days behind despite their later start. The probes could follow a faster, more direct

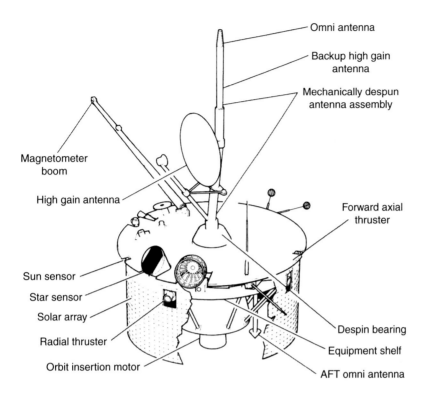

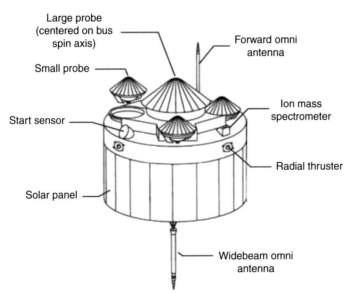

Figure 3.3 (Above) The *Pioneer Venus* orbiter spacecraft, showing the accommodation for the scientific instruments on the upper platform below the communications dish (high-gain antenna). (Below) The multiprobe bus, showing the positioning of the cone-shaped probes on their spring-loaded release mechanisms.

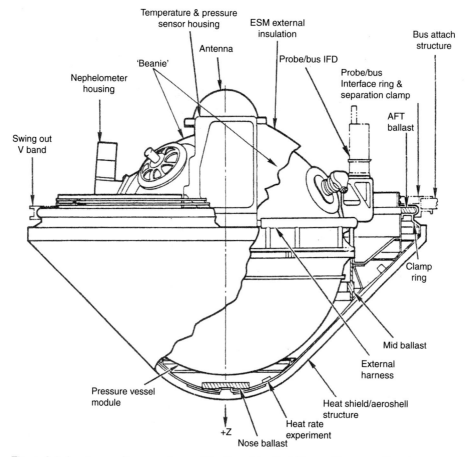

Figure 3.4 A cutaway diagram of one of the three identical *Pioneer Venus* small probes.

trajectory since their velocity relative to Venus on arrival could be higher. The probes plunged straight into the atmosphere, whereas the orbiter had to fire its solid-fuel rocket motor, burning 179 kilograms of fuel, in order to slip into orbit. Figure 3.5 shows the trajectories they followed, and Figure 3.6 the geometry of their deployment at Venus.

As planned, the bus burned up in the upper atmosphere, its two mass spectrometers operating down to an altitude of about 100 kilometres above the surface. The probe instruments were intended to work down to (but not on) the surface, but in the event most of them stopped working simultaneously at a height of about 12 kilometres. A later inquiry showed the likely cause of this to have been the failure of insulating material on one of the external sensors, which caused a short circuit. One of the probes survived its impact with the surface and continued to transmit data for a further hour.

The orbiter functioned throughout its nominal mission of one Venus year (225 Earth days) with the loss of only one instrument, the infrared radiometer, after 72 days. It was impossible to tell what the fault was, except that the instrument worked perfectly for two

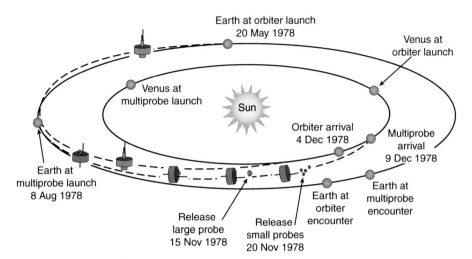

Figure 3.5 The trajectories followed by the *Pioneer Venus* orbiter and multiprobe spacecraft on their journey from Earth to Venus.

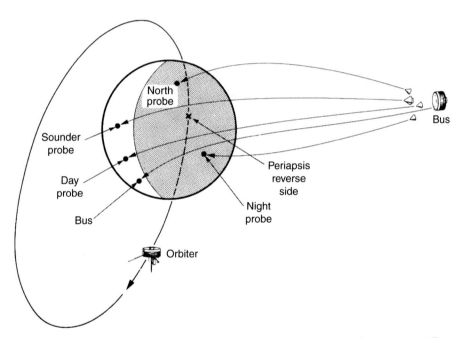

Figure 3.6 The geometry, as viewed from Earth, of the *Pioneer Venus* probe entries on 9 December 1978, showing the orbiter already on station. The large, or 'Sounder', probe landed near the equator, while the three small probes and the bus, which had carried the probes prior to separation, were distributed as shown to characterise daytime, night-time, low latitude and high latitude conditions.

and a half months, then shut down completely, which pointed the finger at an electronics failure. These, to save precious mass, were of a hybrid design that was very advanced for the time, but not as well flight proven as they might have been. However, a huge amount of data covering a lot of the planet had already been obtained, and the scientific return was excellent, including some exciting discoveries, so the team was not too disappointed. The spacecraft, with its surviving payload, went on to obtain valuable data for nearly 14 Earth years, finally entering the atmosphere and burning up during mid-1992.

Vega: the first balloon mission

French space scientists had always been closer to the Russians than any other western country, and they had a dream. This was to send one, or preferably many, 'buoyant stations' to Venus, to better explore the atmosphere than was possible with rapidly descending probes. They designed balloons that could survive the acid in the clouds and float near the one-atmosphere pressure level for many days, to be carried around the planet by the winds (Figure 3.7). Each balloon would have a payload of instruments to measure the temperature and pressure and other key quantities, and would be tracked as it drifted to get data on the wind speed and direction.

The Soviet Union bought in to the idea eventually, and in December 1984 launched two identical probes towards Venus. Named *Vega*,[4] they were designed to fly past Venus and go on to encounter Halley's Comet in March 1985 during its journey around, and close to, the Sun. Each *Vega* was in three parts: the main bus, the only part that would go on to

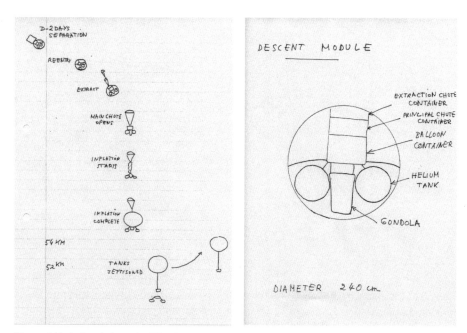

Figure 3.7 A sketch from 1979 by Jacques Blamont, the scientist most active in promoting the French plan to insert floating instrumented platforms into Venus's atmosphere. The plan was eventually realised by the *Vega* mission in 1985.

---

[4] Vega is a contraction of Venera + Gallei, the latter being Russian for Halley.

Halley; a *Venera*-like Venus lander; and the French-designed balloon. As the package approached Venus, the lander and balloon were released on a trajectory that would take them into the atmosphere.

Slowed by parachutes, the balloon and its instruments deployed near the cloud tops, and the descent craft continued down to land on the surface. The landers were similar to the *Veneras* that had gone before, with a comprehensive payload that included a surface analysis package. *Vega 1* failed before landing, but *Vega 2* was able to show that the rocky surface where it touched down was, like previous landing sites, made of solidified volcanic lava. As a sign of the increasing sophistication of the data, the experiment team identified the igneous mineral anorthosite, which is common on Earth. It is also found on the Moon, where it predominates in the lightest-coloured regions.

The balloons were the big first for *Vega*, of course; despite many follow-on proposals, they are still the only buoyant stations so far deployed into Venus's atmosphere. They floated at a level inside the clouds, at an altitude that varied little around 50 kilometres above the surface, gathering meteorological data for about two days.

The designers in France had faced some unique challenges, over and above the obvious difficulties of the balloon surviving while immersed in the concentrated sulphuric acid clouds. It would be exposed to sunlight during the day, which would heat the gas and cause it to expand. Either the fabric would have to be extremely strong and hold a very high pressure, or some of the gas in the balloon would have to be vented into the atmosphere. The latter is undesirable since, if the balloon then crossed the terminator to the nightside, the remaining gas would cool and the balloon would fall. However, it was not possible at the time to make a balloon strong enough to remain totally pressurised on both the dayside and nightside of the planet.

In practice, the solution for this trail-blazing mission would be to accept a limited lifetime, and limit the flight of the balloon to a single terminator crossing. Both *Vega* balloons were deployed near local midnight on Venus, about four days apart. They entered a few degrees either side of the equator, earlier plans for one at the equator and the other near the north pole having been abandoned, mainly because of difficulties in relaying the data. They both covered about 100 degrees of longitude, travelling more than a quarter of the distance around the planet, before contact was lost.

In addition to measuring temperature, pressure and light levels the *Vegas* successfully delivered some information on horizontal and vertical atmospheric dynamics by tracking the position of the balloons as they drifted with the wind. The winds were large by earthly standards, up to 3 metres per second in the vertical direction and 70 metres per second in the horizontal. Bursts of turbulence, lasting an hour or so, were encountered, suggesting convective or stormy regions, but the optical sensors designed to search for lightning flashes drew a blank.

A picture of the planet emerges

The comprehensive coverage of Venus that the *Pioneer* programme achieved with its orbiter and simultaneous multiple probes, including radar maps of the surface, coordinated ground-based measurements and an 'in-house' team of theoreticians (the so-called

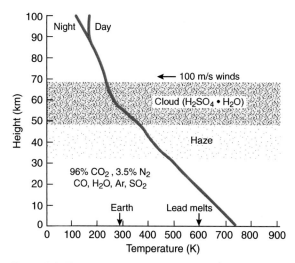

Figure 3.8 The mean temperature structure of Venus's atmosphere was well established by the conclusion of the *Pioneer Venus* and *Vega* missions, as were the composition, circulation and cloud structure, although the variability exhibited by all of these still posed many questions.

interdisciplinary scientists selected by NASA at the same time as the payload instruments), was meant to knit together the puzzling glimpses of Venus from the earlier missions, and so it proved. Of course, it also revealed further new mysteries.

One of the fundamental things *Pioneer Venus* sought to get a grip on was the global structure of the atmosphere, for which the basis is temperature and pressure as a function of height, from the surface to space, over the whole globe (Figure 3.8). The infrared soundings covered the Earthlike regime from the cloud tops to the thermosphere, and the sensors on the probes sounded the dense region below that, down to the surface. The probes obtained simultaneous temperature–pressure data at four sites, which is not a huge sample, but they were well spaced out in latitude and longitude and provided the first estimates of the horizontal gradients in the deep atmosphere. As expected, the gradients are small: an atmosphere as dense as that on Venus below the clouds is in some ways more like an ocean.

All of the probes measured the total radiation from the Sun as a function of depth in the atmosphere. Measurements were made looking upwards and downwards, the latter to see how much sunlight was being reflected upwards by the clouds and the surface. The difference gave the solar heating of the atmosphere versus height and time of day. Similar measurements were made in the long-wave infrared part of the spectrum; from this the cooling could be deduced and the overall energy balance worked out, to put some numbers on the scale of the Venusian 'greenhouse' effect and see if it accounts for the high surface temperature. In addition, local differences in energy deposition are what forces motions in the atmosphere, and data on these could be applied to understanding the atmospheric super-rotation.

Planetwide remote sensing of temperature, pressure and density profiles with the infrared radiometer on the orbiter from the cloud tops to the upper atmosphere revealed

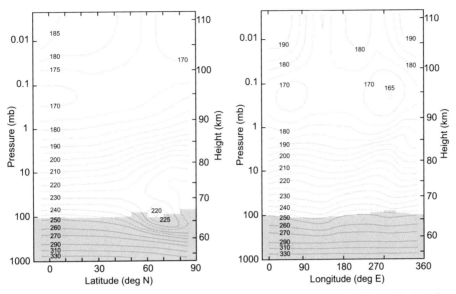

Figure 3.9 Time-averaged temperature fields and mean cloud-top height (dotted line) in the middle atmosphere of Venus from infrared sounding by *Pioneer Venus*. (a) The zonal mean field and (b) the variations around an equatorial belt, both plotted against pressure and height.

several remarkable features for the first time. The first to stand out was the reverse temperature gradient from the equator to the pole, which means that the atmosphere above the pole is warmer than that above the equator, quite a surprise since the Sun is always over the equator on Venus (Figure 3.9, left). Among the global-scale waves seen for the first time were the solar tides, that is, the diurnal cycle of heating of the atmosphere by the Sun, which on Venus has two maxima per day in contrast to the one on Earth (Figure 3.9, right).

A third technique added temperature measurements in the thin upper atmosphere above an altitude of 135 kilometres, the region known as the thermosphere. This could be investigated directly because the highly elliptical orbit made possible measurements of the drag on the spacecraft as it dipped down to altitudes as low as 130 kilometres. Drag data are easily converted to density profiles and thence to temperatures and pressures. The results revealed a surprisingly cold thermosphere, better termed a 'cryosphere', on the nightside, with a remarkably abrupt transition to warmer temperatures on the dayside.

Tracking the probes from Earth yielded vertical profiles of wind velocities at the four entry locations (Figure 3.10). These confirm the rapid winds observed by tracking the movement of the ultraviolet markings in the upper cloud layers, and show how the wind speed declines with height below the clouds.

Above the cloud tops, the winds also decline with height. At first, it is not obvious why they should, since friction generally decreases with density. The answer lies in the temperature gradient produced by the polar warming; this corresponds to a pressure gradient that tends to decelerate the zonal wind. So the counter-intuitive temperature structure of the middle atmosphere turns out to be the result of the dynamical, rather than the

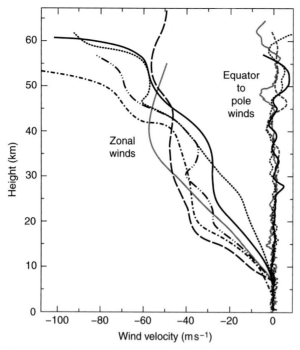

Figure 3.10 Probe tracking provided these vertical profiles of wind speed and direction.

radiative, balance of the region, with both forming essential parts of the four-day circulation.

Observations of the poles on Venus are difficult from Earth, because of Venus's small axial tilt, but easy from a high-inclination orbiter like *Pioneer Venus*. Infrared imaging through the largest telescopes on Earth had managed to see hints of some remarkable temperature and cloud-top structure in the polar regions, apparently a cold collar around the pole, inside which high temperatures could sometimes be glimpsed. This was spectacularly confirmed by images from the spacecraft, looking down from orbit as it passed almost directly above the north pole. The phenomenon, which became known as the 'polar dipole', is a kind of double vortex several thousands of kilometres across rotating around the pole (Plate 8). The polar collar is a cold wavelike disturbance surrounding the dipole that also forms part of the circulation pattern in those regions, marking the boundary between the super-rotating atmosphere at low latitudes and the large polar vortex.

The sounder probe, with its more sophisticated payload, obtained a profile of atmospheric composition and cloud density and microphysics as it passed through each of the layers.[5] The compositional information included the best deuterium to hydrogen (D/H) ratio measurements so far obtained,[6] with its important implications for the history of water on Venus (discussed in Chapter 10). The probe measured the number and extent of

---

[5] Cloud 'microphysics' refers to details such as the size and shape of the individual cloud particles.

[6] Deuterium is heavy hydrogen, with an extra neutron in the nucleus giving it twice the atomic weight of 'ordinary' hydrogen.

each of the cloud layers, the droplet sizes in the clouds, and some details of their composition and optical properties.

The results were not entirely consistent with those from the Russian probes, one of the first indications that the cloud properties vary across the planet. The *Vega 2* particle size spectrometer found very few large particles at altitudes below 55 kilometres, in contrast to results from the *Pioneer* large probe, probably because the probes entered at quite different places and times, and the deepest clouds are the thickest and the patchiest in coverage. The measurements of cloud thickness by the nephelometers on the four *Pioneer Venus* probes differed by less than 20 per cent in the middle cloud,[7] but revealed much larger differences in the particle densities within the lower cloud, as did the instruments aboard *Veneras 9, 10* and *11*. It became clear that not only the cloud opacity, but also the actual type of cloud, vary considerably with height, across the planet, and evolve in time.

The *Pioneer Venus* orbiter found that the top surface of the uppermost cloud layer rises and falls by about 2 kilometres as the local time of day, and so the height of the Sun above the horizon, changes around the equator. This seems to be a complicated function of the photochemical production of H_2SO_4 from H_2O and SO_2 and vertical motions controlled by convective heating, both of which are driven by the daily variation in solar radiation.

Mass spectrometers on the probe and ultraviolet and infrared spectroscopy from the orbiter made refined measurements of the abundances of the various gas species in the lower, mixed atmosphere, and in the upper, diffusively separated region. It was known that sulphur dioxide is an important absorber of ultraviolet radiation, and the new data showed very large fluctuations in its abundance in time and space, so at first it was thought that this gas could account for the markings seen in the clouds at those wavelengths. However, further study shows that the detailed shape of the absorption spectrum of the clouds is not a good match with sulphur dioxide alone; some other absorber must also be present. This is still unidentified, but is probably a solid or liquid component of the clouds (sulphur itself is a strong candidate) rather than another gas, since there are no plausible candidates in the gas phase that would not have been found already by other measurements.

Water vapour measurements are essential, as we struggle to understand why Venus is so dry compared with Earth. However, water is one of the hardest molecules to measure by direct sampling, since it sticks to the walls of the mass spectrometer apparatus, giving readings that are too low, and at the same time outgasses from components of the spacecraft, tending to give readings that are too high. With an unknown combination of these factors at work, the *Pioneer* and *Venera* findings disagreed enormously. Reconciling them is made even more difficult by the possibility that the actual variations in the atmosphere may be quite large, as they are on Earth. A particularly anomalous reading was attributed to one instrument ingesting a liquid cloud particle, sending the reading off scale.

---

[7] Nephelometer means 'cloud measuring instrument'. It works by shining a light from the spacecraft into the nearby atmosphere and measuring the intensity of the light scattered back. This will normally be small if there are no cloud particles present, and large inside a thick cloud. Some nephelometers measure at more than one wavelength, and by comparing these, an estimate of the size of the cloud particles is obtained.

Infrared spectroscopy can be used to measure water vapour, and this has different problems. In those days the technique was not capable of penetrating the clouds, so the data were restricted to the atmosphere above the cloud tops. Here, too, ground-based spectra had shown enormous variations. The *Pioneer* orbiter made a water vapour measurement in its very strong absorption band in the far infrared, and found a localised region with a very high abundance of water downwind of the subsolar point at local noon. This wet patch vanishes as rapidly as it appears as the solar day progresses, and is confined to very low latitudes. Probably the intense convection seen in cloud images in the same region, where the solar heating is greatest, is evaporating cloud droplets and/or bringing up moister air from below.

Aided by the complete angular coverage that it achieved as a result of its spin-stabilised configuration, the *Pioneer Venus* orbiter obtained the first comprehensive maps of the Venus magnetosphere and the solar wind interaction with the planet. This turns out to be quite different from the Earth, the primary reason being that the core of Venus, although probably Earthlike in size and composition, generates little if any global magnetic field. The new data set a very low upper limit on the magnetic moment of Venus; less than one ten-thousandth of Earth's. Why the dynamo process in the two planets should be so different is still a puzzle, and one that we discuss in Chapter 9.

The ways in which the Sun delivers particles and their kinetic energy into the atmospheres of the two siblings are compared in Chapter 8. In both cases the solar wind, a supersonic plasma flow consisting mostly of protons, produces a bow shock when it hits the planet and decelerates. At Earth, the particles are deflected by the magnetic field, while at Venus they impinge directly on the upper atmosphere. The ionopause is the level at which the solar wind dynamic pressure is balanced by the thermal pressure of the exosphere. In this region, atmospheric gases are dissociated and ionised by photon and electron impact and carried away by the plasma flow in a comet-like tail. This can be a major source of net atmospheric mass loss over extended periods of time.

Pioneer mapped the airglow on the dark side of Venus. This is the phenomenon, familiar on Earth, where dim, spectral emissions of visible, ultraviolet and infrared radiation are generated by the chemical recombination of molecular fragments produced in the ionosphere. The new data gave clues to the ion composition, temperature, flows, electron concentration and temperature of the region, high in the upper atmosphere. The satellite also confirmed the earlier detection by *Venera* of radio signals whose characteristics suggested that they probably originate from lightning discharges in the clouds.

By tracking gradual changes and small perturbations in the orbit of *Pioneer Venus* over the course of its 14-year lifetime, it was possible to construct maps of the gravity field of the planet. This, when combined with the radar altimetry results and overall size and mass data, provided the evidence that the interior of Venus has a metallic core and rocky mantle on the same scale as the Earth, and probably with very similar compositions. This makes Venus more Earthlike than Mars or the Moon, not just because of its similar size, but also because the other two have smaller metallic cores as a proportion of their radius.[8]

---

[8] Mercury, on the other hand, is nearly all core. Special conditions apply there, however, because it is so close to the Sun, where the more volatile rocky minerals found it hard to condense when the planets were formed.

There is a basic difference between Venus and Earth in that on Venus there is a strong positive correlation of gravity with topography. The radar altimeter covered most of the surface of the planet and found familiar volcanic and tectonic features such as rift valleys, mountains, continents and volcanoes. When the spacecraft passed over the largest features, its orbit was perturbed, minutely but perceptibly, more than would be expected in Earth orbit. The reason is probably that the outer solid crust on Venus is much thinner than Earth's, so anomalies have a larger effect. However, understanding the details would require more sophisticated mapping, which was to follow a decade or so later.

Chapter 4
Images of the surface

The first glimpses of the surface using radar from Earth

The surface of Venus has long been hidden from human eyes by the thick veil of clouds, leading, as we have seen, to all kinds of speculation about what lies beneath. However, radar can be bounced from the surface and the recorded echoes synthesised into a picture, and today most of our detailed knowledge and mapping of the surface has been obtained in this way[1]. To get good resolution, and to cover the polar regions, the radar equipment needs to be on a spacecraft orbiting Venus.

However, before the first mission to carry radar flew to Venus in 1978, remarkable progress had been made in obtaining pictures of the surface using the same technique all the way from the Earth. This requires a very large dish antenna to transmit and receive the pulses, and those at Goldstone in California and at Arecibo in Puerto Rico were the first to be pressed into service.

The detection on Earth of the first radar reflection from Venus, a breakthrough for any planet, was achieved on 10 March 1961 at the Goldstone station.[2] Two 25-metre diameter antennas about 10 kilometres apart were used, one as the transmitter and the other as the receiver. Venus was then at inferior conjunction, its closest point to the Earth. The fact that the observable disc is dark then is of course not important to the radar, which does not rely on the Sun as a source of signal, but the fact that it is 'only' 40 million kilometres away makes it easier to detect the feeble returned pulse.

The first image in which surface features could be identified, and then only as vague blobs, was obtained at Goldstone during the next conjunction in 1962. The blobs could be timed rotating around the planet and used to get a reliable rotation rate for the planet at

---

[1] More recently, it has been realised that the surface can also be imaged at certain near-infrared wavelengths that, like radar, can also pierce the cloud veil (Plate 10). This technique is discussed in Chapter 6.

[2] A total of five different groups of observers, including one in Russia and one in England, announced the detection of radar echoes from Venus at about this time. Earlier claims, by MIT Lincoln Lab in 1958 and Jodrell Bank in 1959, were considered unreliable as the signal-to-noise ratio was not high enough.

last. The result was a surprise, not so much because it was slow (250 Earth days, with an uncertainty of 50 days[3]) but because it was *retrograde*. If we imagine ourselves suspended in space looking at Earth to our left and Venus to our right, Earth is turning from left to right and Venus from right to left. By some definitions, we would say that the north pole of Venus is at the bottom, not the top, of the globe, although to avoid this sort of confusion the usual convention is to say that north in the Solar System is defined by the rotation of the Sun.

It seemed then, and in fact it still seems, very remarkable that Venus should rotate in the opposite direction to most of the other planets[4], and it raises big questions about angular momentum conservation during the formation processes that produced the entire Solar System. The early radar data also showed that Venus's axis is close to being perpendicular to its orbital plane, with neither pole tilted significantly towards or away from the Sun at any time. Thus, there would be no equivalent to the seasons that are so familiar to Earth-dwellers.

As the radar technique was improved, the blobs began to take shape as identifiable geological constructs (Figure 4.1). If multiple receivers are available, a clever interferometric technique can be used to obtain high spatial resolution, even on distant targets.[5] In this way, features as small as 10 kilometres across had been identified in radar images of both Mars and Venus by mid-1975.

In the images, regions of high radar reflectivity appear bright, indicating variations in the composition, in the surface morphology, or both. In other words, bright areas can appear so either because they are composed of material which is an intrinsically better reflector, or because the surface is smoother than surrounding darker areas, or because the surface is tilted to be more nearly perpendicular to the incoming beam than the local horizontal.

By the time of the 1975 conjunction, large, elevated plateaus – what we would call continents on Earth – and giant volcanoes were being observed on Venus in remarkable detail. Figure 4.2 shows the vast mountain range called Maxwell as it was seen then.[6]

In 1992, high resolution images of Mercury were achieved as well, using the large dish at Goldstone to transmit a pulse of nearly half a million watts. The return signal was received by the Very Large Array, a vast network 20 kilometres wide of receivers at the National Radio Astronomy Laboratory in New Mexico. This achievement led to the discovery of thick deposits of water ice in the shaded craters near the poles on Mercury.

---

[3] The modern value, as determined by the *Magellan* mission discussed later in this chapter, is 243.0185 days, with an uncertainty of about 1.5 minutes, and it is indeed retrograde. Recent measurements by *Venus Express* (Chapter 8) have found that Venus's rotation rate has slowed by 6.5 minutes since the *Magellan* measurement, probably due to angular momentum exchanges between the solid planet and its atmosphere. The Earth shows a similar, but smaller, effect.

[4] Uranus is also strange in this respect, having its axis of rotation within a few degrees of its orbital plane.

[5] An interferometer is a device that combines two or more separate signals from the same source to produce fringe patterns by interference between the two. These can be analysed to give information about the source at much higher resolution (effectively that of a telescope with diameter equal to the separation between the receivers) than the individual signals contain.

[6] Maxwell is the tallest mountain on Venus at a height of 11 kilometres, and also has the distinction of being the only feature anywhere on the planet not named after a female, but for the physicist James Clerk Maxwell. On its relatively rarefied and chilly summit, the pressure is only 60 bars, and the temperature a mere 380°C.

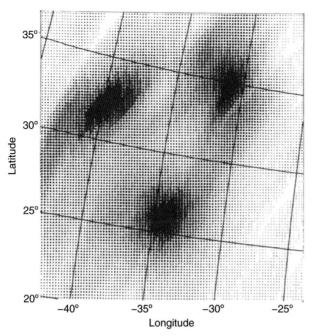

Figure 4.1 An early (*c.*1967) Goldstone radar image of the region on Venus known as Beta (see also Figure 4.8). Alpha (α) and Beta (β) were the first to be identified as prominent regions on the planet, and the names are still used for these two to this day, although Gamma, Delta, etc. have been superseded by classical names.

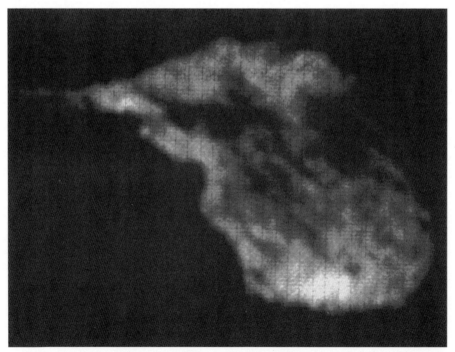

Figure 4.2 The highest mountain range on Venus, Maxwell Montes, seen in radar images of the planet obtained by the Arecibo radio telescope in Puerto Rico. The observations were made during the inferior conjunction, when Venus was at its closest to Earth, in September 1975. Maxwell is about 750 kilometres across in the north–south direction (vertical in the picture) and details down to a scale of perhaps 20 kilometres can be discerned in the image.

Pioneer and *Venera*: early radar maps from orbit

Radar mapping can be accomplished much more readily by placing the transmitter and receiver on a spacecraft. Of course, the antenna that can be carried is much smaller than the Earth-based giant dishes, and there is far less power available for the transmitted pulse, but being so much closer to the target more than compensates.

The first mission to study Venus with radar was the *Pioneer Venus* orbiter, which, with its polar orbit, not only had the advantages of proximity to obtain better spatial resolution and topographic information, but also was able to map parts of Venus's surface that had not been covered before, especially regions near the poles which are always viewed obliquely from Earth. Maps were slowly built up strip by strip as the planet turned slowly beneath the spacecraft, taking a complete Venus year of 243 days to complete. The error in altimetry was less than 200 metres, with a surface footprint as small as 10 kilometres in diameter at closest approach when the spacecraft altitude was about 200 kilometres. This resolution provided a good overall view of the features on about 90 per cent of the surface of the planet, as the map in Plate 9 shows.

The globe of Venus turned out to be almost perfectly round, and most of the surface relatively flat. The squashing down of the poles seen on the Earth is not found on Venus, nor is the double-peaked distribution of surface heights that characterises the lofty continents and deep ocean beds of Earth. Mars is more like Venus in this regard, but has a wider spread of terrain of different heights than that of Venus, which is strongly peaked at the median value (Figure 4.3).

In fact, more than half of Venus's surface is within 500 metres of the mean surface height. This means that an ancient ocean would leave less than 10 per cent of the surface of Venus uncovered, if it had the same depth on average as that we have on modern Earth. Thus the highlands cover only a small fraction of the total surface area of the planet, and that includes the two large continents. Ishtar Terra is about the size of the continental United States, and Aphrodite is about the size of Africa. The highest point on the planet, found in the Maxwell Montes shown in Figure 4.2, lies within Ishtar.

In addition to the continents and giant mountains, the *Pioneer Venus* radar images revealed large-scale tectonic features such as rift valleys and volcanic calderas.[7] However, there were none of the long puckered ridges that occur at the edges of the large, slowly moving plates that are characteristic of what used to be called 'continental drift' on the Earth[8]. This apparent absence of plate tectonics on Venus, when it should have been readily detectable even in the relatively crude *Pioneer Venus* maps, was a surprising and significant difference from Earth, with many implications for the interior and even the climate, as we discuss later.

---

[7] 'Tectonic' features are those that are produced by movements of the crust, resulting in cracking, lifting, shearing and so forth.

[8] The term 'continental drift' is now considered obsolete among geophysicists since it is associated with a theory of the motions that is no longer accepted. 'Plate tectonics' is preferred instead when describing how large slabs of the crust move relative to each other.

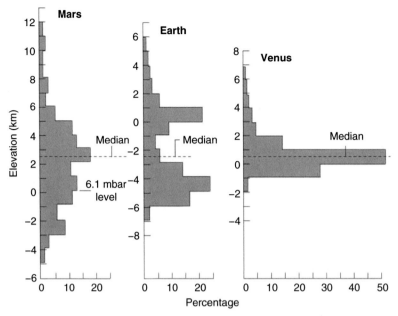

Figure 4.3 The distribution of surface elevation on Venus, Earth and Mars. The sharp peak around the median value for Venus shows that the planet is relatively flat compared with the others.

A few years after the *Pioneer Venus* maps were obtained, comparable quality (although not global coverage) was becoming available from Earth-based radar observations as the method and the technology gradually improved. The extra resolution obtained by the fairly small instrument on the orbiter, with a mass of less than 10 kilograms and only 18 watts of power, less than a domestic light bulb, was important, but still not fine enough for geologists to study the individual structures in the kind of detail they needed to really understand the processes shaping the Venusian surface. For that they would require a synthetic aperture radar (SAR), a much larger, heavier and more sophisticated device.[9] The Russians were developing just such an instrument for their next mission.

Venera 15 was launched on 2 June 1983 and *Venera 16* five days later, entering Venus orbit on 10 and 14 October of the same year. Each carried the very large antenna, data processing electronics and power systems required by a SAR, taking up essentially the entire science payload of the 4-ton craft (Figure 4.4). The 'synthetic aperture' feature involves taking multiple samples of the surface by successive radar pulses as the spacecraft moves along its orbit. These are later synthesised into high-resolution images of the surface, giving a performance equivalent to an even larger stationary antenna. The *Venera* radars obtained a resolution of around 1 or 2 kilometres, more than ten times better than *Pioneer Venus*.

---

[9] A SAR works by repeatedly mapping the surface as the viewpoint changes because of the movement of the satellite in orbit. A sophisticated analysis of the overlapping views allows much higher spatial resolution to be obtained, so that additional details can be seen in the terrain below.

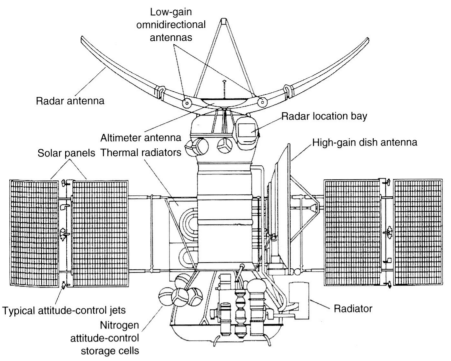

Figure 4.4 In this diagram of the *Venera 15/16* spacecraft, the large dish at the top is the antenna for the radar, while communication with Earth was achieved by the smaller dish on the right side of the main body of the spacecraft.

Each spacecraft was in a nearly polar orbit with a closest approach (periapsis) of about 1,000 kilometres, at a high northern latitude. Apoapsis for the very elliptical orbit was much further away, more than ten times the radius of the planet. This gave an orbital period of 24 hours, which is convenient for mission operations back on Earth.

The two spacecraft were inserted into Venus orbit with their orbital planes a few degrees apart, to make it possible to reimage an area if necessary. In June 1984, Venus was at superior conjunction and passed behind the Sun as seen from the tracking stations during part of the mapping sequence. No communication was possible then, so the *Venera 16* attitude control jets were employed to shift its orbit back by 20 degrees after the conjunction was over, to allow it to map the areas it had missed.

Together, the two *Veneras* imaged approximately a quarter of the total surface area of Venus, including the north polar region, over the eight months of mapping operations. Typically for the time, the Soviets were slow to release the data, which meant they got less attention than they might have done in the West, where NASA was thinking about its own Venus mapping mission. But this too was slow to come, and for nearly a decade the *Venera* images of the surface of Venus were the best available, and were responsible for a number of important discoveries.

For instance, it was found that many of the circular features, that had previously been thought to be meteorite impact craters, were in fact the low dome-shaped volcanoes now known as coronae. The total number of impact craters that have been found on Venus is in fact quite low, partly because the meteors have to be big and strong enough to survive travel through the thick atmosphere, and partly because large areas of the surface are covered with relatively recent lava flows that bury and conceal any craters formed earlier.

Magellan: high-resolution radar images of the surface

NASA began a study in the early 1970s to design a mission called *VOIR* (*Venus Orbiter Imaging Radar*[10]) that could achieve a resolution on the surface about ten times better than *Veneras 15* and *16*. The science study team believed that many key features of the terrain would be revealed by this improvement. However, *VOIR* was cancelled in 1982, at a fairly advanced stage, because of runaway costs that eventually exceeded a billion dollars.

In 1983 the concept was rescued with a smaller, cheaper version called *Venus Radar Mapper*, which dispensed with all of the scientific payload except the radar, and made more use of spare parts and existing designs from other spacecraft. After numerous delays, including the fallout from the 1986 disaster involving the space shuttle *Challenger*, this finally set off from Cape Canaveral in May 1989 under its new name, *Magellan* (Figure 4.6).

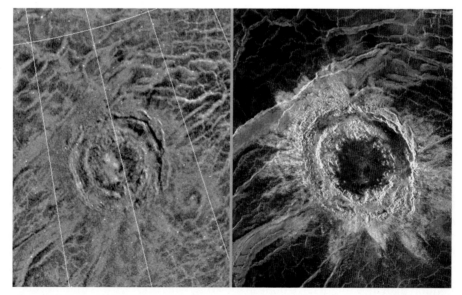

Figure 4.5 A comparison of images of the same crater in Tethys Regio, about 100 kilometres across, from radar data obtained by *Venera 15* (left) and *Magellan* (right).

---

[10] NASA distributed a lapel badge to the study team that was a cut above the usual with an elegant pun. It said 'Nous allons VOIR Venus'. However, they pronounced the mission acronym V-O-I-R.

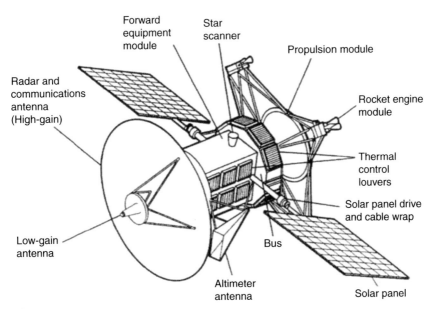

Figure 4.6 The *Magellan* spacecraft. The large dish (3.7 metres across) was used to transmit and receive the radar pulses directed down to the surface of Venus. The measurements were stored, and then later in each orbit the spacecraft was rotated to point towards Earth so that the same dish could be used to relay the data back to the ground station.

Figure 4.7 A *Magellan* map of Aphrodite Terra, the largest 'continent' on Venus. In terms of area, Aphrodite is about the same size as Africa.

Magellan achieved radar imaging of the surface with a resolution of around 100 metres, taking Venus from being the most obscure to one of the best-explored terrains in the Solar System. The composite image of Aphrodite Terra in Figure 4.7 gives some idea of the detail that was obtained, as does the comparison with *Venera 15* in Figure 4.5. It was said at the time of the end of the *Magellan* mission that the high-resolution maps we now had of Venus were better than for some parts of the Earth.

The features first glimpsed as dark patches in the early radar images from Goldstone, were now revealed as extensive areas of high ground containing mountain ranges (see, for example, the image of Beta Regio in Figure 4.8).

Figure 4.8 Beta Regio was one of the first features detected in the earliest radar images obtained at Goldstone (Figure 4.1). Then it was just a group of smudges; now it is seen in the *Magellan* images as consisting of two large volcanic complexes with their associated lava flows. Rhea Mons is the more northerly bright feature, with the more circular Theia Mons to the south.[11] They are about 800 kilometres apart. *Venera 10* landed on one of the dark plains just to the south-east (bottom right in the image) of the mountainous ridge.

---

[11] Theia and Rhea are female Titans in Greek mythology.

Nearly all of the structures on the surface of Venus can be placed in one of three categories, those produced by:

1. volcanic activity,
2. tectonic activity (crustal movements), and
3. impacts of meteorites large and strong enough to reach the surface without breaking and burning up as they speed through the thick atmosphere.

Magellan data revealed an enormous number of objects of obviously volcanic origin on the surface of Venus. These range from massive peaks with lava flows extending hundreds of kilometres, down to much smaller, but still unmistakably volcanic, domes, cracks and vents. *Magellan* scientists say they counted over 1,600 features definitely of volcanic origin, including 168 large volcanoes. Figure 4.9 shows two of the latter fairly close to each other, in an example of how the *Magellan* data can be

Figure 4.9 *Magellan* data processed to give a three-dimensional view of the 1.5 kilometre high volcano Sapas Mons,[12] located in the Atla Regio area of Venus. It is about 400 kilometres across, with a double peak and lava flow patterns extending large distances. In the background is another volcano, Maat Mons,[13] the highest feature in the southern hemisphere and, after Maxwell, the second highest anywhere on Venus.

---

[12] Sapas was the goddess of the Sun in Canaan, an ancient region west of the River Jordan.

[13] Maat was an Egyptian goddess whose image is common in the tombs of the pharaohs, where she is seen wearing an ostrich feather in her headdress as a symbol of truth.

cleverly processed to give a three-dimensional view as if we were flying over the surface in an aeroplane.

Because the coverage of the planet was not complete, and the limited resolution of the pictures made smaller features hard to classify, the actual number of volcanic features of all sizes on the whole of the surface of Venus can only be estimated, but at around one million, the number is impressive. The immensely important role of volcanism in shaping the surface, and controlling the climate, will be covered in later chapters.

Volcanic features are not just the volcanoes themselves but also the vast lava flows which are seen draining from them and filling the plains that cover huge areas of the planet. Some areas appear to have been flooded several times, with fresh flows partially covering the earlier ones. Because of this resurfacing, a lot of the surface of Venus is what geologists call young, with an age of perhaps 'only' 100 million years or so.

The example in Figure 4.10 shows a plain that cracked extensively after it solidified, and which is studded with small raised features. These are probably small volcanoes. Cracks and vents are very common features all across the planet, but more unusual in this case are the fan-shaped streaks behind the small peaks, all oriented in the same direction. These could be the result of ejecta from the vents being carried by the

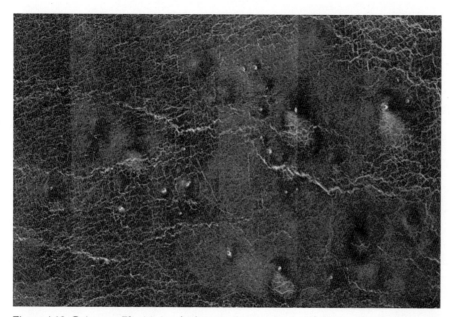

Figure 4.10 Guinevere Planitia is a fairly typical example of a relatively smooth plain on Venus. Note the cracking of the lava that occurred when it solidified, and the bright streaks or fans behind the small raised cones on the plain all aligned in the direction of the prevailing wind.

prevailing wind, or, more likely, the result of the wind scouring the ground and depositing dust long after eruptions ceased. Either way, the wind seems to have blown consistently in one direction for an extended period. Studies of hundreds of streaks have shown they tend to be aligned towards the equator and towards the west, which is consistent with the models of the circulation of the atmosphere discussed in Chapter 13.

Figure 4.11 shows examples of three of the most common types of tectonic features. All of them involve segments of the crust pushing against each other to produce ridge patterns of various sizes and types, including cliffs, ditches and canyons. Sometimes they are seen as complex branching networks, and sometimes they form the spidery radial patterns called *nova*, where deep plumes that carry heat to the surface from the interior are pushing upwards and producing cracking. The *tesserae* are networks of ancient terrain that has been cracked

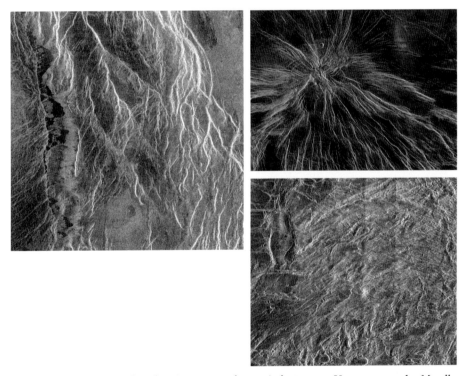

Figure 4.11 Three examples of various types of tectonic features on Venus, as seen by *Magellan*. At left is Devana Chasma (named after a Czech hunting goddess), a rift valley 1,000 kilometres long produced by the crust pulling apart. At top right, radial cracking caused by a convective plume of subsurface lava pushing upwards at the centre produces a nova in the Themis region. At bottom right is an example in Alpha Regio of tesserae, relatively ancient terrain cracked into small plates by the accumulated effects of numerous crustal movements.

and jumbled to produce a rectangular pattern, hence the use of the Latin name for tiles.

Although there are many examples of the crust moving on a fairly local scale, it was widely accepted by the experts on the *Magellan* team that the new data showed no evidence for global plate tectonics on Venus. Despite the higher resolution, there is still no sign of a Venusian version of the characteristic giant mid-ocean ridges at the edge of continental-sized moving plates interacting as they do on the Earth. One view, that we will revisit in Chapter 11, is that we should not expect to find such features on a planet without oceans, since abundant liquid water is required to lubricate the plate movements.

Where there are ridges and other features produced by deformation of the crust on Venus, they seem to be less weathered than similar features on Earth, because they have sharper edges and steeper slopes. This is the case even for structures that otherwise seem to be billions of years old, and at first is surprising given the extreme conditions of temperature and pressure on Venus. However, it is probably the dryness of the surface and atmosphere that counts most when it comes to the preservation of ancient geological features. Apparently the absence of ice and rain, of diurnal and seasonal temperature cycling, and of high winds, means the kind of erosion that occurs on Earth is quite muted on Venus.

However, liquids of some description have flowed on Venus and cut features that look like river valleys (Figure 4.12). It is not completely impossible that this was water, in an earlier, cooler climate, but it is much more likely to have been some type of lava that is very runny and mobile, which formed long flow patterns over an extended period of time in the same way water does on Earth. The rivers often lead into the vast plains, covering most of Venus, and the plains clearly were formed by flooding with lava. Some of the riverbeds are partially obliterated and therefore quite ancient, and a few actually run uphill, presumably showing that there have been alterations to the terrain since they formed. Others look younger and may actually contain liquid that is flowing at the present time. It is not possible to tell from the data we have at present.

Magellan confirmed that there are relatively few impact craters on Venus, compared with expectations based on the cratering of airless bodies such as the Moon and Mercury that were exposed to a similar level of bombardment. The remnants of impacts that we do see in the radar images of Venus, such as the example in Figure 4.13, are more or less uniformly distributed over the surface. A few of them have been modified by lava flows, but most have not. Taking these three lines of evidence together, the most likely explanation is that relatively recent resurfacing means that only the most recently formed craters are on view. This is considered by the specialists who have studied all of the data in detail to be evidence for massive, planetwide lava floods around half a billion years ago, with much more limited volcanic activity since. Whether this is true, and if so whether it is a cyclic or a

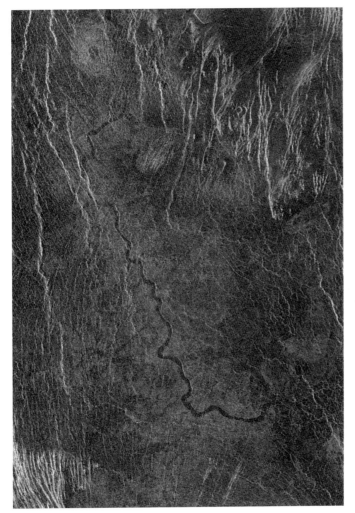

Figure 4.12 An example of a river valley on Venus. This one is about 200 kilometres long, and 2 to 3 kilometres wide. The superficial resemblance to such features on Earth does not, of course, mean that the fluid that cut the channel was water; certain kinds of lava known on Earth would be very fluid under conditions on Venus and thus it seems that something of this kind flowed quite freely there at certain times and places.

one-off characteristic of Venus volcanism, are debatable topics to which we will return in Chapter 11.

The summit of the lofty mountain Maxwell shows an example of another curious phenomenon. Like the riverbeds, this illustrates a characteristic of Venus that is reminiscent of a familiar feature on Earth, and yet must be dramatically different. The

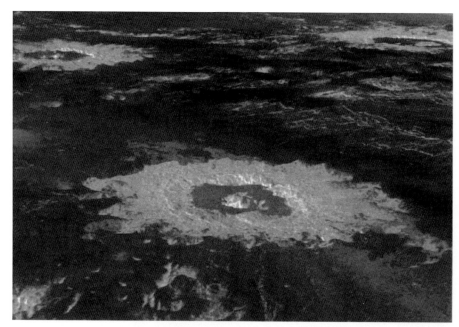

Figure 4.13 Many Venusian impact craters have splash marks around them where debris was thrown out during the impact. This Magellan view captures three examples, all craters 40 to 60 kilometres across, that show this feature clearly: Saskia in the foreground, Danilova to the left and Aglaonice to the right.[14]

bright material in the picture in Figure 4.14 is some substance with an anomalously high radar reflectivity that has apparently condensed from the atmosphere onto the highest ground, like the snowcaps on a mountain on Earth. Nearly all of the surface that is more than about 2 kilometres above the mean height on Venus shows this coating, and in each case it starts at the height where the temperature is the same, 438 degrees centigrade.

The 'snow' also shows up in a different type of image obtained by *Magellan*. In the example in Plate 11, the source is thermal emission from the surface rather than a reflected radar pulse. These maps show less structural detail than radar, but give different information about the nature of the surface, since the signal strength depends on surface emissivity. Emissivity is usually a function mainly of the composition, and is usually high when the reflectivity is low, and vice-versa. The volcano in Plate 11 shows low emissivity

---

[14] Aglaonice was a female astronomer in ancient Greece; Danilova was a Russian ballerina; Saskia was Rembrandt's muse (and is also a dragonslayer in a video game, although that may not be relevant here).

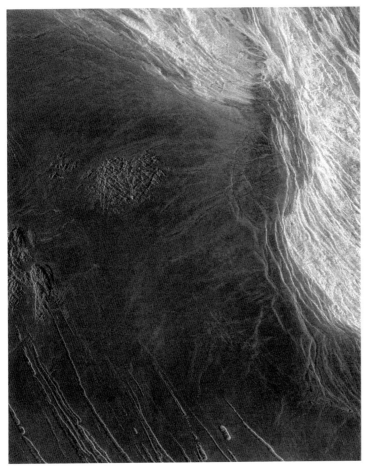

Figure 4.14 A close-up radar image of Maxwell, the largest mountain on Venus, obtained by *Magellan* (compare with Figure 4.2, where this region is at the bottom right). The dark side to the left of the peak is very steep, and features patches of ancient jumbled terrain (tesserae) that apparently escaped when the rest of the face was covered with lava. The side to the right is flatter and higher and covered with 'snow'.

from the radar-bright coating at the summit, in contrast to the high-emissivity lava flows that surround it. The actual values provide a clue when trying to understand what material may be snowing out onto the relatively cool, but still scorching, high ground on Venus.

Chapter 5
The forgotten world

The 'evil twin' syndrome

After *Magellan*, there was a long hiatus in the exploration of Venus by spacecraft that extended from the end of the radar-mapping mission in October 1994 until the arrival of *Venus Express* in April 2006. The campaign to kick-start that European mission relied in no small part on pointing out that our nearest neighbour had become the 'Forgotten Planet', rather as Mars did for a time after the successful landing of the *Viking* surface stations.

However, at the turn of the millennium, interest in comparative planetology was at an all-time high and it might have been logical to focus our available resources on Earth's nearest neighbour and closest twin. The *Venus Express* advocates were also at pains to point out that one of the key comparative aspects was global climate change, with Venus as the ultimate example of a greenhouse-warmed, Earthlike planet.

Looking back, it still seems surprising that such an interesting and accessible target had suddenly taken a back seat in everyone's plans for new missions. One reason, of course, was that the greatest champion of Venus exploration, the Soviet Union, underwent a major political upheaval, one consequence of which was that their scientific space programme became moribund. Where they had once regarded Venus as a 'Russian planet' that showcased their skills and outlook, much as manned landings on the Moon had done for the United States, only a few scientists remained engaged in Venus research and quite a few of those emigrated to Europe and the United States.

Even without the huge political changes that were taking place in the East, Venus studies had a problem. The very success of the *Veneras* and *Vegas*, and of NASA missions such as *Pioneer Venus* and *Magellan*, had painted a picture of our neighbour in the common mind that was complete enough to show that, as a place to land, Venus is not altogether alluring. While scientists exulted in the discovery of such a remarkably massive atmosphere and alarmingly extreme climate, and sought to get to grips with the reason why this second Earth had turned out so different from the first, the popular media tended to focus on the fact that the surface of Venus was uninhabitable and could not be visited, even with robots, if the explorers were to survive more than a very short time.

Time and again, articles appeared, even in astronomy magazines that ought to know better, that stressed not how interesting Venus's curious environment was, but how hellish it seemed, with crushing pressures, baking temperatures and corrosive acid clouds. The contrast with the earlier expectation that Venus would be a steamy, swampy jungle probably teeming with life was a stark one, and the label that said 'Earth's twin' was corrupted by someone into 'Earth's evil twin'. The label stuck and has been regurgitated endlessly ever since.[1] Because planetary missions are expensive, and always need a sound base of support from the public, media and politicians, the image of 'nasty Venus' has been less than helpful. And, of course, the idea that we will never land there, nor will we find life,[2] does take away two important goals for anyone, scientist or lay person.

Meanwhile, the Americans had been distracted away from Venus, too: they had rediscovered the attractions of Mars as a destination. That planet had suffered its own hiatus following the successful, but in some ways disappointing, *Viking* missions of the mid-1970s. NASA had come to a slow realisation that *Viking* had looked in the wrong way for life on Mars, and that the red planet had undergone fascinating climate change from what most believed was an early Earthlike state.[3] They were planning a new series of spacecraft to renew the exploration of Mars, beginning with a large orbiter, *Mars Observer*, in 1991, and leading to rovers and sample return before 2010.

At the same time, NASA had decided it was time to commence the serious exploration of the outer Solar System, focusing first on Jupiter. They had visited the largest planet four times already, with two *Pioneer* and two *Voyager* spacecraft, but these flew quickly past the giant planet and obtained only quite superficial data. It was time for an outer planet orbiter, and probes into the atmosphere, and the closest and best target to begin with was Jupiter. As they built the large and expensive spacecraft, later to be christened *Galileo*, they were also planning its successor, which would orbit Saturn and deliver two probes, one into the planet and the other onto its giant, cloudy satellite Titan. This later became (without the Saturn probe, which was cancelled to save mass and money) the hugely successful *Cassini* project.

In Europe, where a planetary mission capability was just emerging, thanks largely to the participation of European scientists in Soviet landings and the American orbiters and probes (*Pioneer Venus* for the atmosphere, *Magellan* for the surface), Venus was seen as having been 'done', with few really high-priority objectives left. A picture of Venus as an Earthlike body with a varied, rocky surface and a hellish climate had congealed. The unattractive nature of the environment was part of the problem here as well: the public could at least imagine the possibility of indigenous life on Mars, and if not that, human colonies from Earth. Such ideas looked like non-starters on Venus.

These feelings were to predominate for the next two decades, resulting in the absence of any new, dedicated missions to Venus by any nation. At the same time that the size of the scientific community was shrinking in response to this lack of activity, profound questions were slowly being defined about our remaining lack of understanding of Venus and its differences from Earth. Eventually these questions would become sufficiently intriguing to spark a small revival in mission activity.

---

[1] Just googling 'Venus evil twin' produces 1,690,000 hits. See also Figure 5.1.
[2] Or will we? See Chapter 15.
[3] For the full story, see the author's book, *The Scientific Exploration of Mars* (Cambridge University Press, 2009).

Figure 5.1 The 'evil twin' in a BBC report. It is hard to find any report in the mainstream media about Venus that does not use the term.

Unfulfilled objectives

There was no shortage of mission planning during the period when there were no launches. It is always the case that, for every mission flown, there are a dozen or so detailed studies of similar or rival missions that never get beyond the drawing board, or even the final report stage. A few, such as the *Venus Orbiting Imaging Radar* mission discussed in Chapter 4, make it all the way to the end of Phase A, which means experiments and investigators are selected and detailed designs and costs worked out, before they are wastefully cancelled, but this is rare. As long as there were no new flight missions to Venus in the pipeline it would act as a spur to produce more and more studies, each defining exciting objectives and unsolved problems, and designing mission concepts and hardware to address them.

Because *Magellan* had no atmospheric science capability, the planners in the USA had been refining their goals for new missions targeting this area since the end of *Pioneer Venus* in the 1980s. The *Pioneer* team itself pointed out in 1989 that the frequency of all planetary missions was in sharp decline, with 32 launches in the 1960s, 11 in the 1970s, but only two in the 1980s. They attributed this in part to the movement towards very large 'flagship' missions such as *Galileo*, the Jupiter orbiter that was then under development, which absorbed all of the available funds, especially if delayed. The flagships should be augmented with a number of small missions, they said, and proceeded to define a few.

For Venus, they chose to recommend a new large probe that would land on the surface and survive for an hour or so, *Venera*-style. The instruments would follow up on the unsolved problems of atmospheric composition and cloud chemistry by making more sophisticated measurements, including the abundances of noble gases, and the isotopic ratios of the more common elements carbon, hydrogen, oxygen, nitrogen and sulphur. The search for cloud constituents other than sulphuric acid, including the elusive ultraviolet absorber, and the putative large, solid, crystalline particles tentatively detected by the *Pioneer* probe, would continue. NASA Ames Research Center, which had successfully managed the *Pioneer Venus* programme, estimated that it could do this new mission for just US$140 million.

The most authoritative body for defining and recommending strategies for future missions was the Committee for Lunar and Planetary Exploration (COMPLEX). This had access to NASA expertise, but was run independently by the National Academy of Sciences in Washington. In 1987–1988, they reviewed objectives for Venus, again concentrating on atmospheric topics because the *Magellan* surface-mapping mission was already under construction. The papers from those meetings again show a strong focus on the cloud structure, the chemical cycles involving sulphur, water and chlorine, and isotopic ratios, including deuterium to hydrogen with its implications for primordial oceans on Venus. Also highlighted are the unexplained forcing for the high zonal winds in the middle atmosphere, and meteorological phenomena such as planetary-scale waves and lightning. Finally, they called for analysis of the surface materials by landers, including the abundances of oxides and sulphides and other minerals that interact with the atmosphere.

In 1992, the National Research Council, which represents all of the US National Academies (Science, Engineering and Medicine) published the results of a powerful study called *New Frontiers in the Solar System*. The Venus component of its final recommendations was a mission called *Venus In-Situ Explorer*, known as *VISE*, which combined a descent module, a buoyant station and a lander that could drill and analyse a core sample of the surface rock. The payload was designed to study the geochemical cycles in the clouds and near surface, and what was called 'non-equilibrium surface environments'. The latter expression is, in other contexts, often associated with exobiology, but in this case was intended to apply more to understanding such things as how Venus's atmosphere keeps its carbon dioxide, and its high levels of reactive sulphur compounds, without weathering processes converting them to solid sediments. The overarching goals were towards understanding volatiles on planetary bodies, specifically the history of water on Venus, and 'the dramatic differences in planetary evolution especially as manifest in their diverse surface environments'.

NASA hoped for foreign collaboration and cost-sharing for expensive and ambitious missions like *VISE*. By this time, Russia was not a likely candidate and they looked to Japan and Europe. However, Japan liked to work alone on its plans for missions to the Moon, Mars and Venus, and the Europeans disappointed everybody when in 1994 the European Space Agency (ESA) updated *Horizon 2000*, its definitive long-term plan, and failed to mention Venus at all.[4]

Technical studies

All space agencies have programmes of technology development that try to anticipate the future flight programme, in order that a new mission, once approved and funded, might proceed expeditiously without undue delays and cost overruns. Naturally these developments tend to be guided by the scientific studies carried out by the academies and other erudite bodies, including the agencies themselves. For Venus, the pressing need was to invent hardware that could work in extremely high temperature and pressure environments for extended periods. These ranged from pressures of a few atmospheres, at temperatures below 100 degrees centigrade, for aerobots sampling the clouds and winds, to 100 atmospheres at temperatures well above the melting points of the softer metals such as lead and tin for landers to drill and analyse the surface.

Somewhere in between were the aircraft or 'submarines' that could float in the dense air above the surface and navigate around the topography in order to investigate and map the diversity of the surface regimes, from the lava beds to the mountain tops. They might encounter conditions that are extreme even by Venus standards, for example by flying through erupting volcanic plumes. They might also land at various interesting places in order to sample the crust or study rock formations close up. In any case they would have to have a long lifetime – certainly weeks rather than days, as achieved so far by *Vega*, and then only in the safest part of the atmosphere, that is, near the cloud tops where conditions are Earthlike.

These engineering studies threw up problems that form another large part of the reason why Venus exploration stalled: most of the major goals identified by the scientists required

---

[4] Actually, there was one sentence in the 'Conclusions' chapter, that stated laconically (under the subheading 'Other planets') that 'the atmosphere of Venus remains largely unexplored'.

technology that the engineers could not deliver with the current state of the art. In many cases – high-temperature mechanisms and electronics, power sources, refrigerators – even long looks forward into the future do not yet offer a lot of promise. Even so, the research community came up with a string of ingenious proposals for projects that could be carried out with existing or feasible engineering, and were affordable. The following examples of 'phantom missions' that could have happened but didn't are just a few chosen from a shelf full of dusty proposals in the author's office. The hard work that went into the concepts and designs they contain is often lost and later re-invented, since the documentation with the details is usually unpublished and often considered secret by its proponents, who don't want to give their ideas away to rival proposers. The process by which real missions eventually fly is a long, tortuous and wasteful one.

Phantom missions

Hadley and *VADIR*

One objective for Venus that could be addressed with existing and inexpensive instrumentation was the remarkably fast atmospheric circulation, including the super-rotating winds and the giant double vortices at the poles. In 1982, ESA was looking for a new project and one of the six candidates it evaluated was *Hadley*, a simple spin-stabilised Venus orbiter. The name derived from the Hadley regime, the term for a global equator-to-pole overturning of the atmosphere that was found on Earth by the eponymous meteorologist,[5] and which is now found to dominate the circulation on Venus as well. With the first dedicated, systematic coverage the mission would examine this in detail. It would also investigate the drive for the four-day 'super-rotation', and (best of all, the proposers said) would obtain the high-resolution coverage in space and time that would reveal the secrets of the curious 'polar dipole' phenomenon discovered by *Pioneer Venus*.

The dipole, the proposal alleged, deserved our attention as a major Solar System phenomenon, intimately connected with the general circulation of Venus's atmosphere, and one which after years of trying we are still struggling to explain. Like the Jovian Great Red Spot, the absence of viable theories which can be tested, or in this case any really concrete theory at all, leaves us uncomfortably in doubt as to our basic ability to understand even the rudimentary features of planetary atmospheric circulations. New observations were needed. In particular, infrared images with better resolution and coverage than *Pioneer*. These would not be difficult to acquire with the latest technology: a high-resolution multispectral imager, a deep-atmosphere microwave radiometer and a high-resolution Fabry-Perot spectrometer to obtain wind velocities from Doppler shift measurements.

The *Hadley* concept was derived within the framework of a possible British-led mission to Venus with a simple and inexpensive spacecraft. The UK funding bodies (the country did not have a space agency until 2010), in a very temporary fit of optimism, had called for

---

[5] George Hadley (1685–1768) was actually a lawyer by profession, but like many others in the seventeenth century he was a renaissance scholar with wide and varied interests.

such proposals to boost the nation's industrial base. Realistically, it was always clear that such a flight opportunity was very unlikely to materialise in the foreseeable future. The first attempt to win European support also failed, being in competition with the successful *Cassini* Saturn Orbiter–Titan Probe mission. But the team persisted and presented updated versions at every opportunity.

The new version, now called by the less parochially English name *Venus Atmosphere Dynamics Imaging Radiometer (VADIR)*,[6] was even less complicated than the original concept, having only two spectral channels in the infrared 'window' near 12 microns wavelength. *Pioneer Venus* had shown how the structure of the dipole could be detected and recorded, but this time there was to be high enough spatial and time resolution to see what was actually going on. The concept was for a spinning spacecraft in a near-circular orbit about 300 kilometres above the cloud tops, with the spin axis parallel to the surface of the planet and to the orbital path (Figure 5.2).

For a spin rate of 5 revolutions per minute, the sub-spacecraft point advances across the cloud tops by nearly 100 kilometres each time it spins. A linear array of 20 detectors, each sized for a 5×5 square kilometre footprint on the planet, would then provide slightly overlapping coverage from each swathe. In this way, a detailed image of the dipole could be built up each orbit, or approximately 50 times each dipole rotation period, which *Pioneer Venus* had found was about 72 hours. Spectral resolution would be sacrificed to give a high sensitivity to contrasts in the cloud-top temperature even with the rapid scan. From the images, motion pictures could be constructed and used to study in detail the structure and variability of the dipole.

The *Pioneer* pictures, which over the pole were at much lower spatial and temporal resolution, had nevertheless revealed sufficient detail in the dipole structure to imply that there were many small-scale cloud features which could be tracked to determine

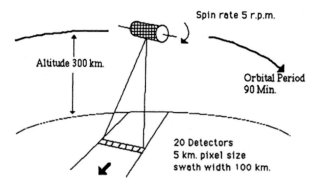

Figure 5.2 An original sketch of the concept for *VADIR*, a very simple and inexpensive mission with ambitious goals.

---

[6] One French member of a key ESA committee actually told me that he had voted against going forward with 'adley' because he didn't like the name. *VADIR* reminded people of the *Star Wars* films that were popular at the time and was more universally acceptable.

cloud-top wind fields within the vortex. In this way, it was hoped that the true nature of this complex and fascinating phenomenon, a major puzzle in the field of planetary atmospheric dynamics, would eventually be revealed. Although the mission never happened, elements of *Hadley/VADIR* were incorporated into *Venus Express*, and the latter did succeed in revealing many of the secrets of the 'dipole', as described in Chapters 8 and 13.

VESAT, VALOR and *VESPER*

While *VADIR* was seeking unsuccessfully to infiltrate the collective consciousness of the European space science community, the larger scene in the USA was far from dormant. Scientists in universities and government laboratories there advanced proposals for Venus missions at every opportunity, sometimes several at the same time, in competition with each other. The one that came closest to success, *VESAT* (*Venus Environmental Satellite*), was similar in many ways to the European proposal, emphasising simplicity and low cost with a single small orbiting spacecraft that focused on a limited number of goals, mostly related to the basic circulation of the atmosphere.

VESAT promised to address the mysterious forces that power and sustain the high winds by 'globally mapping the cloud-tracked windfield at high spatial and temporal resolutions over an extended range of altitudes, time, and solar lighting conditions'. Thus it would 'delineate the roles of solar-induced thermal tides, travelling waves, eddies and gravity waves'[7] and 'measure the horizontal momentum and heat transports due to the mean eddy circulations'. Secondary objectives were to study cloud formation, dissipation and chemistry, and to study the surface emissivity, surface temperature, and the temperature profile in the lowest 10 kilometres of altitude above the surface. Volcanic activity should be 'readily observable' via temporal and spatial variations in the surface thermal flux and by localised variations in water vapour, sulphur dioxide, carbonyl sulphide and other volcanically generated species.

With just three instruments covering selected parts of the ultraviolet, near- and mid-infrared spectrum, a small spacecraft weighing just over 200 kilograms and an inexpensive launch vehicle (*Delta II*) could deliver new results in a key area of Venus science for the kind of modest cost NASA was looking for at the time (this was 1996). But there were so few mission opportunities that the opposition, coming from all areas of Solar System science, was stiff. *VESAT* was proposed again in 2000 in a more ambitious form, now adding a microwave instrument for sounding the sub-cloud atmosphere, and promising 3-D movies of the polar dipoles, the Hadley cells and the super-rotation. It failed once more, however.

---

[7] Thermal tides are temperature changes driven by the change in solar heating during the course of a day, and are not necessarily a simple function of local time. Travelling waves are like the ripples in a pond when the surface is disturbed. Eddies are chaotic, often (apparently at least) random disturbances due to instabilities. Gravity waves are oscillations between the downward pull of gravity and a restoring force, usually buoyancy in the atmosphere, but a bungee jumper at the end of his drop is another example.

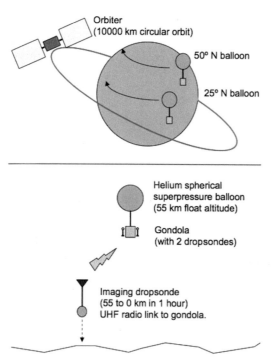

Figure 5.3 The *Venus Atmospheric Long-Duration Observatory for in-situ Research (VALOR)* mission concept.

In 2006, some of the same team abandoned the orbiter concept and came up with *VALOR*, the *Venus Aerostatic-Lift Observatories for in-situ Research*. As the name implies, it now offered floating balloon-borne platforms to address the noble gas and isotope abundances, and the cloud physics and chemistry by making direct-sampling measurements using a gas chromatograph/mass spectrometer and other instruments. Winds could also be measured by tracking the drifting balloons, although only for two days, for the same reasons that had limited the lifetime of the *Vega* balloons. *VALOR* was an advance on the earlier mission mainly because it had a much more sophisticated payload, and because two platforms would be deployed, one near the equator and the other at a high latitude where it could sample the polar vortex (Figures 5.3 and 5.4).

After *VALOR* failed to progress, despite being highly rated, the team took yet another different tack for the next opportunity to propose, which came at the turn of the millennium. In 2000, *VALOR* now stood for *Venus Atmospheric Long-Duration Observatory for in-situ Research*. Progress in balloon design meant that a single super-pressure balloon could survive inside the clouds at an altitude of 55 kilometres for two months rather than just two days. This would give it time to circle the globe of Venus several times, covering a range of latitudes, and do all this with a larger payload than previously planned. The flight across the boundary zone from outside to inside the polar vortex was expected to be particularly exciting and insightful.

However, once again the mission was not selected by NASA's panels, which chose the *Dawn* mission to the asteroid belt instead. *Dawn* recently completed a survey of the minor

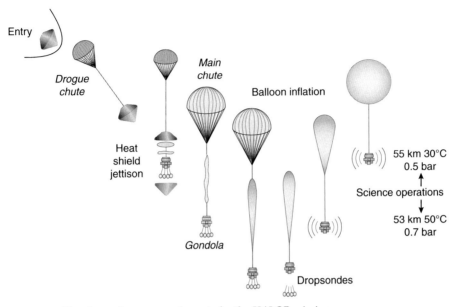

Figure 5.4 The planned sequence of events for the *VALOR* mission.

planet Vesta, and is currently travelling towards its larger sibling Ceres, where it arrives in 2015.

While the Jet Propulsion Laboratory pursued *VESAT* and *VALOR*, another team based at NASA's Goddard Space Flight Center on the east coast was developing *VESPER*.[8] This was another orbiter, targeting atmospheric chemistry and dynamics like *VADIR* and *VESAT*, but with the defining feature that the main instrument on board was to have been a submillimetre heterodyne spectrometer capable of very sensitive measurements of rare chemical species like oxygen and chlorine oxide, and of the water family including the rare species hydrogen dioxide, HO_2. Like *VALOR* it failed, but came back in improved form in 2010, only to lose again.

Goddard scientists also offered an *in situ* Venus probe in 2010 called *VISAx*. This was not a floating station like *VALOR*, but rather a direct landing with a pressurised probe in the same style as the early *Veneras*, or like *VISE*. Of course, scientific instruments had improved greatly since the 1970s, and the planned payload was much more sophisticated than the Soviets had been able to deploy. It included descent imagers to obtain high-resolution imaging of the terrain in one of the ancient highland regions, and a high-resolution tunable-diode laser spectrometer to augment a sensitive mass spectrometer to detect some of the more elusive minor constituents of the atmosphere.

---

[8] At first sight *VESPER* is not an acronym, but the old Latin name for the Evening Star. However, the mission's proposers succumbed to the prevailing trend and devised another excruciatingly artificial construct, 'VEnus Sounder for Planetary ExploRation'.

Broken dreams

Europe finally made the breakthrough and had a real mission under way with *Venus Express* in 2005, and Japan followed with *Akatsuki* at about the same time (approval is a stepwise procedure in Japan, so the start of a mission is not dated precisely). In the USA, however, every proposal for a new flight to Venus failed.

The best the world's leading space nation could do at this time, for the planet it had been the first to reach, was to formalise the lobbying effort and focus future studies. The Venus Exploration Analysis Group (VEXAG) was established by NASA in July 2005, its stated goal being to identify scientific priorities and strategy for exploration of Venus. The organisation has five focus groups:

– Goals and Objectives
– Completed Missions and Data Analysis
– International Venus Exploration
– Technology Development and Laboratory Measurements
– Young Scholars

and an overall chairperson, and works by actively soliciting input from the scientific community and discussing it in regular meetings. VEXAG reports its findings and provides input to NASA, but is not supposed to lobby its recommendations. We will return to the VEXAG vision for future Venus exploration in Chapter 17.

Chapter 6
Earth-based astronomy delivers a breakthrough

In the mid-1980s, Venus exploration by spacecraft received a big boost from what seemed at the time a rather unlikely source: Earth-based telescopes in observatories on mountaintops. This was unexpected because the key discoveries that could be made from 50 million miles away, with our atmosphere in the way, were thought by most of us to be essentially in the past. Close-in polar orbiters, entry probes and landers ruled almost everyone's thoughts – those who thought about Venus at all – as the twentieth century moved towards a close. The ground-based observers were, in any case, more preoccupied with distant galaxies and cosmology, and nowadays are generally reluctant to devote expensive telescope time to poking around in what they see as their own backyard. Also, of course, there was the usual endless rivalry for funds between planetary and deep-space astrophysicists, and between space scientists and traditional astronomers, to be taken into account.

The breakthrough happened in Australia. Astronomers at the Anglo-Australian telescope in New South Wales had developed a new near-infrared imaging spectrometer optimised for imaging extended objects such as star formation clusters and the nearby planets of the Solar System. Venus was prominent in the sky, with a bright crescent showing and most of the disc dark, when the observers turned the big 3.9-metre aperture telescope in its direction in June 1983. Examining the data, they were surprised to find that the nightside was not in fact dark everywhere in the near infrared spectrum, but instead ghostly bright features were present at a few wavelengths just longer than visible light (Figure 6.1).

This finding mirrored the discovery of the ashen light nearly 400 years earlier, except that now sensitive cooled infrared detectors were making the observations. At the same time, as throughout history, some amateur astronomers were still reporting ashen light observations made with the naked eye. Although at first the observers attributed the origin of the nightside features to scattered sunlight from the dayside, or possibly electrical phenomena in the atmosphere, theoretical modelling studies soon demonstrated that they are produced when thermal radiation from the hot lower atmosphere escapes to space through the thick planetwide cloud layers (Figure 6.2). The atmosphere is transparent enough for this to happen only within narrow spectral 'windows' that were found to exist

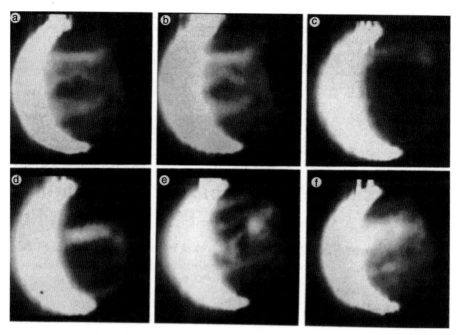

Figure 6.1 Allen and Crawford's original images of Venus showing nightside emissions. The bright crescent is the sunlit side of Venus.

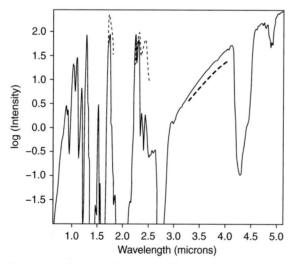

Figure 6.2 The dashed lines are Allen and Crawford's measurements of the spectrum (brightness vs. wavelength) of the infrared windows on the nightside of Venus, compared here to theoretical spectra computed by Lucas Kamp.

between the strong molecular absorption bands of carbon dioxide and water vapour in the 0.9 to 2.5 micron wavelength region. This occurs all over the planet, but can only be detected on the nightside where it is not overwhelmed by the much brighter solar flux reflected from the clouds.

New windows on Venus

Once the source of the nightside emission had been identified, it was quickly realised that its discovery, and the existence of the transparent windows, provided a powerful new technique for studying Venus. Since then, near-infrared imaging and high-resolution spectroscopic observations of this emission by ground-based and spacecraft instruments have been used to investigate the surface and lower atmosphere, regions that used to be inaccessible except to entry probes and radar.

Among the measurements that have been made are particle sizes and concentrations deep inside the cloud layers, winds within the middle and lower cloud decks, and the abundances of several important trace gases, including water vapour, chlorine compounds, carbon monoxide, sulphur dioxide and carbonyl sulphide, in some cases right down to the surface. New information has been gleaned about the near-surface temperature structure, and the deuterium-to-hydrogen ratio with its clues about an early ocean.

Such is its importance that it is surprising at first to think that the existence of the near-infrared windows had not been predicted or observed earlier, especially in view of the ashen light controversy, now centuries old. The reason is that no one had made the connection between two key factors that contribute to the transparency of the Venus atmosphere at these wavelengths.

First, concentrated sulphuric acid cloud droplets scatter almost conservatively in the near infrared, meaning they can redirect photons but do so with very little probability of absorption. In contrast to this, the absorption is very strong at wavelengths longer than about 2.5 microns. So, although photons emitted from gases in the deep atmosphere must encounter many particles in their path through the cloud layers to space, at short wavelengths these encounters predominantly take the form of changes in direction through scattering, and eventually the photons diffuse upwards into space where we can observe them. The Earth's water clouds do not behave like this, but rather absorb quite strongly.

Secondly, it had not been realised how low the absorption by atmospheric gases is, even in the dense lower atmosphere, in the spectral regions between strong carbon dioxide and water vapour absorption bands. This absorption is dominated by the edges or 'wings' of the strong spectral lines in the bands, and calculations made in those days predicted that these would add up to a substantial opacity, even though it would be much smaller than in the band centres.

Provoked by the search for an explanation of the Venus emissions, it was found that the absorption in these wings is substantially less than the calculations had predicted. Theoreticians traditionally used an approximate formula to calculate the complex effects of molecular motion and collisions on the spectral line shapes, and it was found that these simplifications generally work reasonably well for Earth's atmosphere. It turned out, however, that they cannot be extrapolated to the long path lengths and high pressures and temperatures on Venus without introducing large errors.

The fact that the spectral lines do not have strong wings after all means that the continuum absorption between the numerous bands is much weaker than a calculation based on the traditional line shape would predict.[1] Indeed, no complete theory for the line shape exists that predicts from first principles the small absorption in the wings that is observed, and it is difficult to obtain the combination of high pressures and long path lengths required to study it experimentally in the laboratory. The results from Venus therefore not only explained the windows, but also contributed to our understanding of the forces present during the collisions of individual molecules under conditions of high temperature and pressure.

Soon, improved models of the upwelling radiation in the near-infrared windows reproduced the conservative scattering in the planetwide cloud cover, and calculated the fraction that escapes to space. At longer wavelengths the properties of the liquid sulphuric acid droplets are different and they absorb strongly, blocking any further windows in the spectrum of the gaseous part of the atmosphere. Images and spectra taken at long infrared wavelengths show only the faint, nearly spatially uniform thermal emission from the cloud tops at around -30 centigrade. So the calculations agreed with the observations at all wavelengths and could be used as an analysis tool.

Meanwhile, the observers were making progress, too. Much better imaging was possible later in 1983 when the planet presented only a narrow crescent but was still far enough from the Sun to be observed, a condition which occurs every 19 months and lasts for only about two weeks. Time-stepped images taken during September 1983 showed that the near-infrared features moved from east to west, in the direction of the cloud-top super-rotation, but circling the planet in not four but more like five and a half days. Thus, although the bright and dark near-infrared features superficially resembled the ultraviolet markings seen at the cloud tops, they were quite different in detailed shape, and moving more slowly. Their contrast in brightness was also much greater. The reason had to be that the infrared features have their origin deeper in the atmosphere where the winds are less strong, and are produced by thicker clouds.

The leaking 'greenhouse'

The windows allow energy from the surface and lower atmosphere of Venus to radiate directly to space, and the resulting cooling must be a factor in the debate about how the high surface temperature is maintained. It was confirmed at the time of *Pioneer Venus* that the lower atmosphere is heated primarily by solar radiation, with about 1 part in 50 of the energy falling on the planet from the Sun penetrating downwards through the clouds

---

[1] 'Traditional' here refers to the Lorentz pressure-broadened line shape, with a Doppler core calculated using the Voigt formulation, all details that non-scientists can ignore. Alternatively, the basic details of atmospheric radiation and its propagation are discussed in *Radiation and Climate* by I.M. Vardavas and F.W. Taylor (Oxford University Press, 2011).

to the surface. This has to be balanced somehow by the removal of energy, mainly by re-emission to space at longer wavelengths in the infrared.

The high surface temperature is a consequence of the fact that the loss process is quite inefficient, due to the high optical thickness of the overlying atmosphere at most infrared wavelengths. The energy deposited in the lower atmosphere, mostly at low and mid-latitudes, cannot radiate away to space to anything like the extent it does on Earth. Instead, most of it is transported horizontally and vertically by advection of warm air to cooler regions, and finally radiated to space from atmospheric levels near the cloud tops.

This sink for the energy from the surface passing through the atmospheric heat engine is short-circuited to some extent by the flux directly to space in the near-infrared windows. The fact of their existence must tend to cool the surface down relative to what had been expected before they were discovered. However, the windows occupy only a small fraction of the total spectrum, and this is at wavelengths where the energy transmitted is small compared with the mid- and long-infrared, even when the relatively high temperatures of the emitting layers are taken into account. Calculations of the net effect, integrated over wavelength, find that the flux escaping from Venus via the near-infrared windows is in fact less than 1 part in 1,000 of the total at other wavelengths, enough to lower the equilibrium temperature of the surface of Venus by less than 1 degree.

Mapping the deep cloud structure and dynamics

Most ultraviolet and thermal infrared wavelengths do not penetrate to or from the region below the cloud tops, and microwave and radio observations generally do not detect the clouds at all. As a result, except for isolated sampling by entry probes, there was very little data on the global structure of the cloud layers and their variability until it became possible to exploit the near-infrared windows. This could have happened more than a decade earlier than it actually did. The *Pioneer Venus* Orbiter Infrared Radiometer had a wide-band channel for measuring the reflected solar flux, and a narrow one centred at 2 microns in a strong carbon dioxide band, intended for the detection of high clouds. The instrument duly made high-spatial resolution maps of the flux from the near-infrared windows, but only on the dayside where they are swamped by the much brighter reflected sunlight. The experiment could in principle have discovered the window emission as early as 1979, but no nightside data were collected since they were not expected to show any features. In retrospect, this was a major blunder.

At any point in time the intensity of the radiation coming from Venus at wavelengths within the spectral windows changes a great deal with location on the planet, showing that the optical thickness of the cloud deck is variable in space and time. This finding implies that the lower atmosphere is very active meteorologically, one of the most exciting, and in some ways perplexing, consequences of the discovery of the near-infrared windows. Most of the structure seen in images such as those in Figure 6.1 originates in the main cloud deck less than 50 kilometres above the surface, which

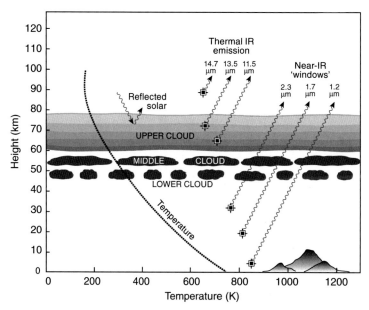

Figure 6.3 A cartoon showing how the emission from Venus originates in the window spectra, with various wavelengths probing different depths.

contains most of the mass of the whole cloud ensemble. It lies well above the atmospheric layers which emit most of the near-infrared flux. In other words, we are seeing the cloud deck as a cold, perforated screen, backlit by the glow from the hot atmosphere below (Figure 6.3).

The remarkable variability of the cloud opacity in a single map, and the evolution with time of the features seen in successive maps, suggests high winds and vigorous convective and wave activity occurring in the troposphere as well as at the cloud tops at the base of the stratosphere. Various features sometimes described as bars, bands and ovals have been seen in the clouds, suggesting a complex meteorological regime at these levels on Venus (Figure 6.4).

These new findings of large-scale, but localised weather within the clouds add complexity, but also potentially insight, to the long-standing struggle to understand the global super-rotation, the phenomenon in which the Venusian atmosphere rotates westward 50 to 60 times faster than the solid body of the planet. As we have seen, this was first described by tracking features in ultraviolet images, but the ultraviolet markings are formed near the cloud tops and may have little to do with the processes that apparently transport the momentum to drive the winds from the planet's surface to these altitudes. The deeper infrared patterns may be more useful once we understand them better.

Tracking the *Pioneer Venus, Venera* and *Vega* entry probes as they descended through the atmosphere showed that the winds blow from east to west at all levels, with wind speeds that decrease steadily with altitude. However, probe measurements are too limited in spatial and temporal sampling to discriminate between

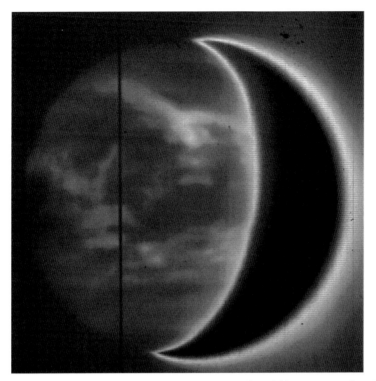

Figure 6.4 A photograph of Venus through an infrared filter at a wavelength of 2.3 micron, obtained by Mark Bullock and Eliot Young at NASA's infrared telescope facility on Mauna Kea, Hawaii, in May 2004. The dark crescent is actually the bright, sunlit side of Venus, which appears dark because it drives the detectors off-scale, while the ghostly bright features on the nightside are much dimmer in reality than they seem here because the contrast in the image has been stretched.

short-lived features due to atmospheric waves and the average wind field. They also lack the accuracy needed to describe the interesting but much smaller winds near the surface, or the weak but important polewards winds at any level. As a result, we are still struggling to understand the role which momentum transport by the meridional circulation and vertically propagating waves may play in the maintenance of the super-rotation. The deep-atmosphere probe afforded by near-infrared spectroscopy might offer the possibility of a more comprehensive global description of dynamics in the middle and lower cloud regions to complement the probe and balloon data, but would need better resolution in space and time than would ever be possible observing from the Earth.

Composition measurements in the deep atmosphere

At high spectral resolution, such as the example in Figue 6.5, the spectra in the windows reveal the absorption lines of many interesting species, including water vapour, carbon

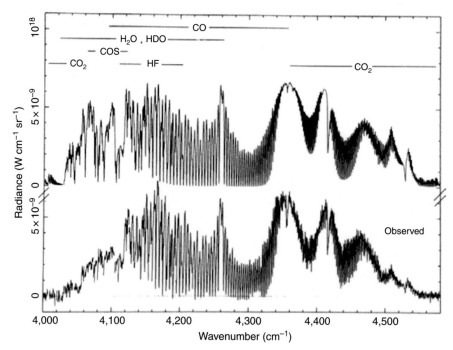

Figure 6.5 A high-resolution ground-based spectrum of the nightside of Venus in the 2.3 micron spectral 'window', compared to a theoretical calculation in which the amounts of the various absorbers (shown by the bars at the top) are varied until the two match.

monoxide, and the sulphur-containing gases emitted by volcanoes. The strengths and shapes of these lines are related to the amount of each gas that is present in the atmosphere, so the abundances can be determined by calculating the spectrum from first principles and finding the amounts that give a good fit,[2] also shown in Figure 6.5.

Because different wavelengths and windows probe different atmospheric regions in Venus's deep atmosphere, information can be retrieved on the species concentration profile as a function of height. The table in Figure 6.6 shows a summary of the results that have been obtained from these and other measurements (including entry probes), and Figure 6.7 expresses these as profiles, produced by making smoothly interpolated averages of the various telescopic and spacecraft measurements.

Water vapour

Water vapour bands are detectable in at least four different spectral windows, each one sensitive to a different range of heights in the atmosphere, depending on the strength of

---

[2] See *Radiation and Climate*. Figure 6.5 follows a convention common amongst spectroscopists in which the horizontal axis is wavenumber, defined as the reciprocal of wavelength, and measured in cm^{-1} (hence 2.3 μm is the same as 4348 cm^{-1}).

Species	Venus	Earth	Climate significance
Carbon dioxide	.96	.0003	Major greenhouse gas
Nitrogen	.035	.770	Similar total amounts
Argon	.00007	.0093	Evolutionary clues
Neon	0.000005	0.000018	Evolutionary clues
Water vapour	30 parts per million	~.01	Volcanic, cloud, greenhouse
Heavy water (HDO)	3 parts per million	~ 1 part per million	Early ocean
Sulphur dioxide	150 parts per million	.2 parts per billion	Volcanic, cloud, greenhouse
Carbonyl sulphide	4 parts per million	0.5 parts per billion	Volcanic, cloud
Carbon monoxide	.00004	.00000012	Deep circulation
Hydrogen chloride	0.5 parts per million	trace	Volcanic
Hydrogen fluoride	0.005 parts per million	trace	Volcanic
Atomic oxygen	trace	trace	High circulation, escape processes
Hydroxyl	trace	trace	High circulation, escape processes
Atomic hydrogen	trace	trace	Escape processes

Figure 6.6 Composition of the atmospheres of Venus and Earth, as fractional abundances except where parts per million or parts per billion are stated.

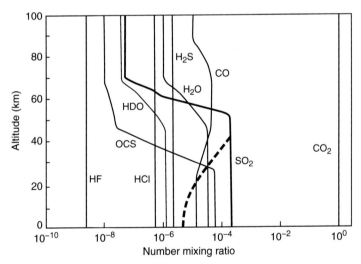

Figure 6.7 A summary of the available measurements of the composition of Venus's atmosphere as a function of height above the surface. The dotted line is the disputed result for sulphur dioxide from the *Vega* probes.

the absorption. The observed emission originates from deeper layers as wavelength decreases, from the surface to just below the lowest cloud, so a crude profile can be determined. If the disc is spatially resolved as well, some limited information on the horizontal variability of the water vapour profile is available, although only at latitudes near the equator if Earth-based telescopes are used.

Different observers at various sites generally agreed that all of the windows were best fitted with a water mixing ratio that appears to be approximately constant both horizontally and vertically, in the lowest 30 kilometres at least, with a value between 20 and 50 parts per million. One campaign found different H_2O abundances according to the brightness of the region, with mixing ratios near 40 parts per million derived from the dark spot spectra, while the spectra of an anomalously bright spot indicated H_2O mixing ratios as high as 200 parts per million. We still do not know whether this was an error of some kind, or an unusual weather phenomenon on Venus, or perhaps a large volcanic eruption contributing additional water in its plume.

Overall the net evidence from remote observations is that the atmosphere of Venus is much dryer than was indicated by most of the *in situ* probe measurements. The reason for the discrepancy is probably the difficulty in getting rid of all the water contamination in the instrument before launch. Water has an unusually high affinity for almost everything, even the metal walls of the mass spectrometer, and it is notoriously difficult to remove even by prolonged baking in a vacuum.

The presence of spectral lines of both normal and heavy water (H_2O and HDO) in the 2.3 micron window can be exploited to obtain the abundances of both species beneath the clouds. The deuterium-to-hydrogen (D/H) ratio from Earth-based studies is consistent with the analysis of *Pioneer Venus* orbiter and probe data at more than 100 times the value found in Earth's oceans and atmosphere. The enrichment on Venus relative to Earth probably results from photodissociation of normal and heavy water molecules in the high atmosphere, and subsequent hydrogen loss to space, with higher rates of escape for the lighter isotope. If so, it must have occurred on a grand scale, and can be interpreted as evidence for a primordial ocean that has since vanished. However, it could be just the steady state corresponding to a balance between loss by escape and supply by comets and volcanic outgassing, since the values for these rates are still essentially unknown. We will return to this question in Chapter 10.

Carbon monoxide

The (2–0) band of CO is a prominent feature[3] in near-infrared spectra of Venus, located near the centre of the 2.3-micron window. The analyses of the first low-resolution observations seemed to indicate carbon monoxide abundances considerably lower than those measured by the gas chromatograph instrument on the *Pioneer Venus* probe, which came up with 30 parts per million at 50 kilometres and the same 12 kilometres higher, falling to 20 parts per million at 22 kilometres. Later, the new high-resolution spectra and a more sophisticated database for the overlapping lines of carbon dioxide found a monoxide mixing ratio and vertical gradient that was in agreement with the *Pioneer Venus* profile, and this appears in Figure 6.7.

---

[3] 2–0 means that the band is made up of lines that correspond to transitions to the ground state from the second excited vibrational level of the CO molecule. Transitions between non-adjacent energy levels produce a so-called 'overtone' band.

Hydrogen chloride and fluoride

The simple, but relatively rare, molecules hydrogen chloride (HCl) and hydrogen fluoride (HF) were first detected at the cloud tops of Venus in 1967 by ground-based observers. High resolution near infrared spectroscopy of reflected light from the dayside yielded mixing ratios of 0.6 parts per million for HCl and just 5 parts per billion for HF. These abundances were below the sensitivity levels of the instruments on the *Pioneer* and *Venera* probes, so they could not be checked *in situ*.

Once lines of both species had been measured in window spectra, 20 years later, spectroscopic analyses of the nightside near-infrared emission could provide the first measurements of the abundances of chlorine and fluorine *below* the clouds as well. In fact, they found concentrations that were essentially the same as those above the clouds, indicating that these gases are well mixed in the atmosphere. This constancy might also suggest that the observed abundances are equilibrium values rather than controlled by episodes of volcanism, although volcanoes are likely to have been their original source. Geochemists have emphasised that these are reactive gases that are likely to be in equilibrium with alkaline rocks on the surface, and the buffering effect could act quickly enough to ensure a near-constant abundance in the upper atmosphere regardless of source activity.

Sulphur compounds

Spectroscopy of carbonyl sulphide (OCS) shows, on analysis, that its mixing ratio falls sharply with altitude, from around 15 parts per million in the lower troposphere to much smaller amounts higher up (Figure 6.7). There is some evidence that this decrease in OCS is matched by an increase in CO of about the same amount, in agreement with chemical theory in which OCS is destroyed by reaction with sulphur trioxide (SO_3) to yield CO above about 30 kilometres. The current best estimate for the OCS mixing ratio near the surface seems to be consistent with the equilibrium abundance calculated from chemical weathering involving pyrite (FeS_2), suggesting that this, and probably the other common iron sulphides as well, is present in significant amounts on Venus's surface.

The sulphur dioxide (SO_2) abundance beneath the cloud decks was first measured by the gas chromatographs carried by the *Pioneer Venus* and *Venera 11/12* probes in 1978. The values were roughly consistent with each other at around 150 parts per million. This is a very large concentration for this gas, which reacts easily with many common minerals at the surface and should be removed quite rapidly from the atmosphere. The high values suggest a steady supply of fresh amounts of SO_2 from active volcanoes.

However, an analysis of the ultraviolet spectra recorded by the *Vega 1* and *2* probes in June 1985 yielded a different SO_2 profile, one that decreased strongly downwards below the cloud decks, and reached just 20 parts per million at 12 kilometres above the surface. This result is unique and controversial, although the experimenters swear by it to this day, and it needs to be repeated sometime soon. If it turns out to be correct, it has major implications for the question of whether SO_2 is in fact in equilibrium with minerals at the surface, and whether Venus is currently the hotbed of active volcanism that most of the other evidence suggests it is.

More recent deep-atmosphere SO_2 mixing ratios derived from nightside spectral window observations made using telescopes in Australia and Hawaii are consistent, within the uncertainties, with those measured by the *Venera* and *Pioneer Venus* probes from 1978 to 1985. The *Vega* experimenters point out that the Earth-based measurements are sensitive mostly to the 35–45 kilometres altitude region, however, whereas the sharp decline that *Vega* found is below that (see Figure 6.7).

The telescopic observers also found no spatial variation in tropospheric SO_2 in spectra recorded in 1989 and 1991 at different latitudes around the equator. This stability contrasts with the apparent massive decrease observed at the cloud tops from ultraviolet spectroscopy, from *Pioneer Venus* and elsewhere, over the period 1978–1988. Since it is not matched by changes in the lower atmosphere, some argue that the cloud-top decrease may indicate long-term changes in the circulation of the upper atmosphere, affecting the composition there, rather than a manifestation of a major volcanic eruption. However, the data sets do not match each other well enough in space and time to be sure one way or the other. The controversy about Venus volcanism rages on, and we pick it up again in Chapter 11.

Chapter 7
Can't stop now: *Galileo* and *Cassini* fly past Venus

Earth-based spectroscopic studies of Venus flourished for a while as planetary astronomers exploited the newly discovered infrared windows. In particular, when these were combined with the data from the earlier space missions, a more complete picture of the atmospheric environment on Venus began to appear.

The success of the telescopic observers served to emphasise the strong argument for taking the powerful technique of spectral imaging in the 'windows' to Venus on the close and versatile platform offered by a spacecraft. As we have seen, none of the *Mariner* or *Venera* spacecraft had this capability, and we *Pioneer Venus* investigators naïvely failed to make and exploit a major discovery by not using its near-infrared capability on the nightside to detect the emission from the deep atmosphere.

By 1989, NASA was preparing the *Galileo* Jupiter orbiter spacecraft for launch. *Galileo* was to reach Jupiter by means of close flybys of Venus and Earth, and would reach Venus in February 1990 (Figure 7.1). As it happened, the Jupiter orbiter carried an instrument perfectly suited for observing Venus, the Near Infrared Mapping Spectrometer (NIMS). Those of us on the *Galileo* team who were also interested in Venus science quickly calculated that NIMS could achieve a spatial resolution on Earth's neighbour that was far better than the Earth-based near-infrared images. Also, it had a spectral range that covered all of the known and predicted windows.

NIMS was designed to map ices and minerals on Jupiter's family of icy satellites and *Galileo* had not been intended to visit Venus at all. That it did so was a result of the accident in January 1986 to the *Challenger* space shuttle, which led to the fleet being grounded for nearly three years. *Galileo* had been scheduled to launch on *Challenger*'s sister ship *Atlantis* using a liquid-fuelled Centaur upper stage. The disaster meant a less powerful solid-fuelled booster had to be used and in addition Venus's gravitational assistance was required to reach Jupiter. This would lengthen the journey from about two to over six years. However, it's an ill wind that blows nobody any good; in this case the detour to Venus offered the chance to make the first close-up application of a technique that was already proving so useful in observations made from the Earth.

NASA initially resisted the suggestion that NIMS should be activated at Venus; it was costly, risky and above all it wasn't in the plan. However, they relented, and NIMS had

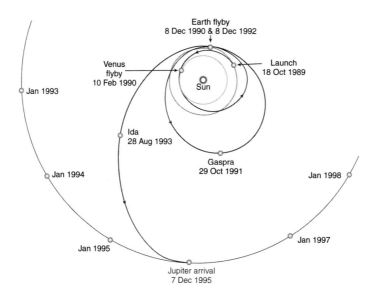

Figure 7.1 *Galileo*'s voyage involved several trips around the Sun, an encounter with Venus and two with Earth, to obtain a gravitational 'slingshot' out to Jupiter. Later, *Cassini* would do something similar in order to get to Saturn using much less fuel than a direct flight would require.

time to make two complete multispectral maps of the planet as it accelerated past Venus on 10 February 1990. The results were spectacular.

Deep cloud structure

Plate 7 shows, on the right, the dark side of Venus mapped in one of the most prominent of the near-infrared windows at 2.3 microns. The clouds are seen illuminated from behind by thermal emission from the hot lower atmosphere and surface. The contrasts are due to the variable opacity of the deepest cloud deck, which contains dark regions of thick cloud separated by relatively clear regions that appear bright in the image. This was confirmation of what the ground-based images had already revealed: that the cloud veil on Venus is not the thick, uniform mass that it had appeared to be for so long.

The *Galileo* infrared maps showed intensity variations of about a factor of 20 between the brightest and the darkest features seen across the disc of Venus, and achieved a spatial resolution of about 25 kilometres when at *Galileo*'s closest point to the planet. The highest observable latitudes are dark and featureless in the near-infrared, indicating relatively high cloud opacity. These dark clouds are at the same high latitudes around the pole as the 'cold collar' detected by the *Pioneer Venus* and ground-based mid-infrared observers.

The bright bands seen at mid-latitudes contain high-contrast cloud morphology that speaks of intense weather activity, featuring wide streaks and large, cumulus-like clumps. Exactly what is going on in meteorological terms is difficult to say from the brief snapshot that results from a flyby, but we could tell that the deep atmosphere of Venus must be a very active place, contrary to expectations at the time. When examining the wavelength dependence of the cloud opacity in different regions of the disc, the *Galileo* team found an asymmetry in the cloud types seen in the two hemispheres. This too seemed odd, since we expect Venus to show a high degree of symmetry between north and south.

Digging deeper, the spectral behaviour of the clouds across the globe seemed to put them in no less than five different categories. It is hard to tell from this data alone what these might signify. The most likely parameter to vary is the particle size distribution, and sizes can be found that would fit the observations, but then we have to wonder what processes are producing the discrete sizes in different places. It could be the result of differences in the vertical velocity, affecting cloud formation, or something linked to different types of volcanism on the surface, as we discuss in Chapter 11.

Because *Galileo* spent less than two days close to Venus, the NIMS observations give no information about the longer-term evolution of global structure in the middle and lower clouds. However, ground-based observations taken over several years revealed a variety of cloud features with a range of sizes and lifetimes. The most remarkable of these was a large, very opaque, patch of cloud that extended about halfway around the planet at low latitudes and survived for at least six weeks.

This feature circulated around the planet fast enough to suggest it was part of the middle cloud region rather than the lowest layer. The ground-based observations taken in January and February of 1990 showed that the most transparent part of this cloud pattern was on the nightside on 10 February, when the *Galileo* observations were taken. This is fortunate because there would have been much less detail to observe two days later, when the thick part of the cloud would have covered the disc with an optical thickness that was so high that little or no near-infrared radiation escaped through any part of it.

Ashen light revisited

As already noted, the patterns of bright emission from the nightside in certain narrow wavelength bands in the near infrared (Figures 6.1 and 6.4) clearly have similarities with the historic reports of ashen light in the visible discussed in Chapter 1. Could they have a common origin? The infrared features seem to be brighter, and are always present, whereas the visible ashen light is notoriously fickle, reported as being sometimes conspicuous and sometimes absent. That could be due to the fact that the optical observers were working near the limits of detection, which clearly they were. The temperature at which the eye sees an object start to glow dull red is generally quoted as being about 1,000 degrees on the Kelvin scale. Venus's surface is more than 200 degrees cooler than this, but for someone with good eyesight in a completely dark room, this may be enough.[1] If so, and we imagine ourselves standing on the surface of Venus at midnight and looking around, we would see the rocks around us glowing dimly in the gloom.

---

[1] This is, admittedly, arguable, as there do not seem to have been any proper experiments done. It may depend on the individual, as some eyes are more sensitive to the longer visible wavelengths than others, and can see a short way into what for most of us would be the infrared. This could explain why some otherwise good observational astronomers say they can never see the ashen light.

The window in the visible part of the spectrum, in which the early observers saw the ashen light, was unfortunately just beyond the short end of the wavelength range of NIMS. If NIMS had observed in the visible, it would probably have recorded images similar to the many sketches that exist, such as that in Figure 1.6. However, it does not automatically follow from success with instruments using cooled quantum detectors that the human eye, a much more modest device in terms of sensitivity, could detect the same levels of emission from the planet using an Earth-based telescope.

When it comes to making a calculation to test the hypothesis that the sophisticated Jupiter orbiter instrument was observing the same phenomenon that a monk in Bologna saw in 1694, the biggest uncertainty is a lack of detailed, quantitative data about the performance of the eye. For an appearance on *The Sky at Night* I estimated how many photons are required to produce a recognisable response and how this varies across the visible wavelength range.[2] With the assumption of a 10 per cent efficiency at a peak wavelength of half a micron, and a Gaussian fall-off to either side of this, the calculation leads to approximate equality with the calculated flux from Venus when viewed through a 10-inch telescope. Pretty convincing, I thought, although the response from the host of the show, Patrick Moore, was non-committal.

Several amateur astronomers have reported regular observations of the ashen light through instruments with this sort of modest aperture. If the response of the eye is assumed, more optimistically, to be 10 per cent everywhere across the visible range, the signal-to-noise ratio increases to about 500. On the basis of this, it might not be unreasonable to expect a fully dark-adjusted observer to be able to see the glow from the surface of Venus under good viewing conditions.

So, in summary, perhaps the famous ashen light is in fact thermal emission from the surface of the planet. The *Galileo* results and radiative transfer models show that light from the surface of Venus, which is hot enough to glow a faint dull red, can propagate through the clouds. At visible and near-infrared wavelengths, the sulphuric acid aerosols, which make up the bulk of the Venusian clouds, are very conservative scatterers, with very little absorption when compared with terrestrial water clouds. The many reports of the colour of the nightside of Venus as being dull red or rusty lend further support to this idea. The observed patchiness is due to the large-scale structure of clouds in the deep atmosphere.

Atmospheric composition

We have seen that the near-infrared spectra provide information on the mixing ratio, horizontal distribution and variability below the clouds, of water vapour, carbon monoxide, hydrogen chloride, hydrogen fluoride, sulphur dioxide and carbonyl sulphide. The NIMS data made it possible to look at these again with higher spatial resolution (but lower spectral resolution, remembering that the instrument was designed primarily to study the broad spectral features of the mineral and ices on the satellites of Jupiter) from close to the planet. It

---

[2] *The Sky at Night* is a long-running popular astronomy programme on BBC TV in the UK, which Patrick Moore hosted from its inception on 24 April 1957 to his death in 2012.

also had the advantage of being able to observe in more of the spectral windows, including the ones denied to the Earth-based observer by the opacity of our own atmosphere.

The NIMS spectra in the windows yielded a water vapour abundance similar to that inferred from ground-based observations, that is about 30 parts per million. Even at the spatial resolution of typically about 100 kilometres, no horizontal variation in excess of 20 per cent was found anywhere across the disc, although the *Galileo* trajectory meant the pass was near Venus's equator and did not allow high latitudes to be covered.

The same constancy was not found for carbon monoxide, however. The principal source of this gas is the dissociation of carbon dioxide high in the upper atmosphere by solar ultraviolet radiation, the process known as photolysis. The highest values of the mixing ratio of carbon monoxide, nearly 10,000 parts per million (i.e. 1 per cent), are found in the dissociation zone. The main sink is in the clouds, where the CO gets caught up in the vigorous photochemical reactions that produce the cloud droplets, and some measurements indicate that the mixing ratio of CO falls to a few parts per million. CO is also probably destroyed by reactions at the surface although we rely on theory for this since the relevant measurements are yet to be obtained. One of the reactions that is expected to be important for removing CO is that with iron sulphide (pyrites), oxidising the mineral and releasing hydrogen sulphide. Reactions involving other sulphur and chlorine compounds will also oxidise CO back to CO_2. Without this recycling, the whole atmosphere would gradually convert into the monoxide.

Below the clouds, the mixing ratio measured by *Pioneer Venus* is about 30 parts per million. The analysis of the CO absorption feature in the set of about 500 *Galileo* NIMS spectra confirms this typical value, but also showed a definite latitudinal trend, with an increase between the equator and 65°N (Figure 7.2) and a hint that it falls off again polewards of the peak. It was harder to say at the time whether a similar enhancement existed in the southern polar region, because the trajectory of the spacecraft favoured observations of the northern hemisphere of Venus and very little high-latitude data was obtained in the south. High latitudes are also inaccessible to Earth-based observers due to the small inclination of Venus's axis of rotation.

Why should there be any carbon monoxide at all below the clouds, where there is no obvious source, only powerful sinks above and below? A possible explanation for the relatively high mean mixing ratio in the lower atmosphere might be the emission of CO by active volcanoes. This could also possibly explain the latitudinal gradient, if there is enough active volcanism on Venus to make a noticeable contribution to the mixing ratio on the global scale.

It might be telling that the biggest mountain range on Venus, the Maxwell Montes, is approximately below the carbon monoxide maximum at about 65°N. For a simple calculation, suppose we assume that the Maxwell range contains active volcanoes that emit as much gas as all of the volcanoes on Earth combined, about 300 million tons per year. Nearly all of this is carbon dioxide, but some of the monoxide is usually present, the proportion depending on how oxidising the environment is deep inside the volcano. Venus may or may not be less oxidising and more volcanically active than the Earth; if we assume that *all* of the carbonic gases emitted are in the form of CO, this will give us an extreme upper limit estimate.

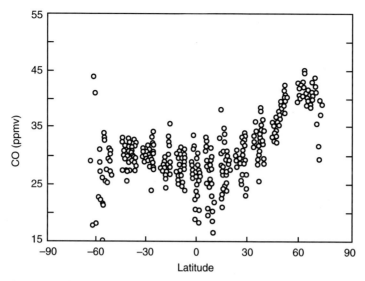

Figure 7.2 The distribution of carbon monoxide with latitude in the lower atmosphere, inferred from observations near 2.3 microns by the near-infrared spectrometer on *Galileo*.

The amount of CO required to produce the anomaly observed by *Galileo*, for a mean mixing ratio of 30 parts per million in the total mass of Venus's atmosphere, is about ten exagrams (that's 10^{16} kilograms). The observed enhancement is one-third of the mean value, over about one-tenth of the surface area, which would require about 300 trillion tons of carbon monoxide. This would take more, probably much more, than 1,000 years at terrestrial emission rates, far longer than the time the CO plume would take to dissipate and be chemically converted, which would be measured in weeks or months.

A better theory would be that carbon monoxide is carried from the upper atmosphere to the lower atmosphere, but then we need to explain how the upper-atmospheric CO is transported through the abundance minimum in the clouds. There is an obvious candidate, since we observe what must be large amounts of mesospheric air descending in the northern polar vortex. This follows since the mass of air transported to high latitudes by the Hadley circulation of Venus's middle atmosphere must be recycled downwards at the pole, as shown in Figure 7.3. The angular momentum that it carries is what produces the vortex in the first place. Infrared observations by the various polar-orbiting satellites do show definite evidence for the depression of the clouds around the poles, which indicates downward motion on a scale that is not present elsewhere on Venus. The observations give us reasonable estimates of the area over which descending motion takes place, but little clue as to the mean downward velocity or, therefore, the rate of mass flow.

If we turn the question around and ask what rate of descent would be required to provide the required transportation of carbon monoxide to match the observations, the answer is about a millimetre per second. This is a very modest vertical velocity for what appears to be quite a vigorous vortex, and drives us to favour transport from the upper atmosphere as the mechanism producing the large (relative to Earth, which has about

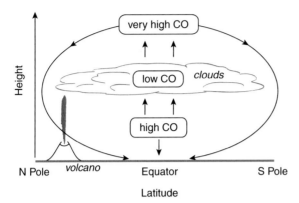

Figure 7.3 A cartoon of the two candidate mechanisms – highly localised volcanism and the Hadley circulation – that might be producing the carbon monoxide distribution observed by *Galileo*.

300 times less) mixing ratio in the troposphere, and also the gradient with latitude observed by *Galileo*.

An obvious way to test this hypothesis would be to see if there is similar, symmetric behaviour in the southern hemisphere. The latitudinal coverage from *Galileo* was not adequate to say one way or the other; nor could the test ever be possible for Earth-based observers. Instead, a high-inclination orbiter, such as *Venus Express*, is required. This eventually addressed the question in 2007, nearly 20 years after *Galileo* reached Venus, as we relate in Chapter 8.

Atmospheric dynamics and meteorology

During an Earth-based observing campaign conducted to support the *Galileo* flyby, the near-infrared markings were tracked to produce a current picture of horizontal winds in the cloud decks. The middle cloud was dominated by the long-lived feature already mentioned, covering most of one side of the planet. The reason for saying it is part of the middle cloud is the rotation period of five and a half days, indicating equatorial velocities near 80 metres per second, corresponding to the middle cloud altitude. There were some short-lived small-scale features that had rotation periods of seven and a half days, placing them in the lower cloud. The latitudinal variations of the zonal motions of both cloud decks resemble solid-body rotation, with no systematic north–south component greater than about half a metre per second in the middle cloud, or about 7 metres per second in the lower cloud. No sign of the Hadley circulation, in other words, but it is expected to be slow and could have been below the detection threshold. Also, cloud-tracked winds can be unreliable at low speeds, as the clouds form and dissipate as well as moving with the wind.

The two NIMS images acquired during the *Galileo* flyby provided better data on the motions of small near-infrared features, since their resolution was about ten times better than that of the best ground-based results. In contrast to the ground-based observations of solid-body rotation, the NIMS results show an increase in the westward wind velocity with latitude. The team acknowledged, however, that there is a difficulty in separating latitude from height variations. The observations might be explained if the lower latitudes were dominated by features in the more slowly moving lower cloud layer, while the high latitude regions were dominated by features in the faster upper cloud region.

The really important result from the same data was that they revealed slow poleward motions in both hemispheres, of the order of a few metres per second. Even if the measurement uncertainties were of the same magnitude, the presence of the meridional component of the Hadley circulation was confirmed.

Temperature maps of the surface

Between them, the spaceborne and ground-based observations made during the *Galileo* flyby detected some additional spectral windows at wavelengths only slightly longer than visible that were being observed for the first time. Their existence had been predicted by theoretical modelling studies, but the terrestrial atmosphere obstructs them when viewed from Earth. In this they differ from those that the astronomers first discovered and exploited, which was only possible because they coincide with a window in Earth's atmosphere as well.

The models also made the exciting prediction that the first three of these new windows, the ones at the shortest wavelengths, would allow viewing right down to the ground and could be used to map the Venusian surface, and they did. In fact, the total column opacity of the atmosphere is low enough that the surface contributes about 60 per cent of the emission in the 1.18-micron and 40 per cent in the 1.1-micron windows. In the 1.05-micron window, just beyond the visible band in the very near infrared, more than 95 per cent of the radiation measured at the spacecraft comes from the surface of Venus.

Plate 10 (top) shows a NIMS image at 1.05 microns, and compares it with the *Pioneer Venus* radar map of the surface. Just a glance at the two placed side by side reveals a strong correlation between high features measured by radar altimetry and cold features in the infrared. The interpretation is obvious: the most prominent, that is the highest, surface features are colder and emit less thermal radiation, as a result of the natural fall-off of temperature with height in the atmosphere. For the largest and highest features, the contrast is large enough to stand out despite the fact that it is being viewed through the clouds.

A lofty feature such as Maxwell Montes, for example, at 11 kilometres above the mean surface height, is more than 100 degrees colder than the surrounding plains and shows up very clearly in the infrared maps made from space. In order to make the infrared and radar observations directly comparable, both are presented in Plate 10 as heights using the same colour scale. To achieve this, the 1.05 micron brightness was converted to a height assuming a vertical temperature gradient of 10 degrees per kilometre, which is the dry adiabatic lapse rate for conditions on Venus and which probably applies, with only small variations, to the lower atmosphere everywhere on the planet.[3] The NIMS surface data did

---

[3] The dry adiabatic lapse rate is the vertical temperature gradient predicted from basic thermodynamic formulae for air containing negligible amounts of water vapour.

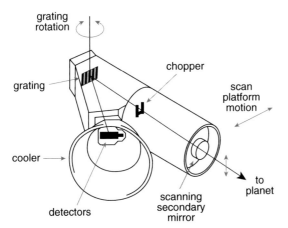

Figure 7.4 VIMS and NIMS were very similar instruments, consisting of a telescope with a built-in scan mirror, a grating spectrometer and an array of cooled detectors. The cooling was by a passive radiator plate, facing cold space and surrounded by a shield to exclude sunlight or planetary infrared radiation. Note that the shield points at right angles to the telescope, since obviously the latter has to point at the planet.

not have enough quality to allow much significant scientific analysis, beyond the very important discovery that the surface was accessible to optical imaging from orbit after all.

Observation of Venus by *Cassini*/VIMS

A decade later, on 24 June 1999, another large imaging spectrometer visited Venus while en route to one of the outer planets. This time the destination was Saturn. The *Cassini* spacecraft flew past Venus and obtained the first observation of Venus's surface emission features at wavelengths shorter than 1 micron with the Visual-Infrared Mapping Spectrometer (VIMS). This instrument was built and flown to obtain spectral images of the atmospheres, rings and satellites in the Saturn system using a design similar to but more advanced than NIMS (Figure 7.4).

The *Cassini* investigators had less luck than *Galileo* in persuading NASA to operate their instrument during the Venus flyby. This time the mission managers decided that the hot environment around Venus meant that the aperture door for the radiator that cools the detectors had to remain closed throughout the encounter. The infrared detectors do not work without cooling, so this prevented the acquisition of the longer wavelength spectra that would have again probed the clouds and measured trace chemical abundances in the middle and lower parts of the atmosphere.

However, the visual channel used an uncooled detector, similar to that in an ordinary phone camera, so this was working normally. It could observe the nightside at wavelengths shorter than 1.05 microns, where calculations of the expected spectrum predicted that two additional windows, sensitive to surface emissions, should occur at 0.85 and 0.90 microns. Although they are not as transparent as the 1.05 micron feature used by NIMS, these are potentially sensitive to surface composition. The *Cassini* investigators

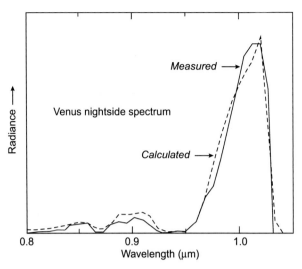

Figure 7.5 The nightside spectrum of Venus observed by *Cassini* VIMS, compared to a computed theoretical spectrum. In addition to the broad peak at 1.01 microns, VIMS detected emission features in smaller transmission windows at 0.85 and 0.90 microns, in agreement with theory.

determined that the identification and mapping of hematitite, pyroxene, olivine and other ferric and ferrous materials in surface basalts might be possible from observations at these wavelengths. *Cassini*'s fast flyby offered just one 12-minute observing pass over Venus, during which time VIMS obtained 64 spectra of the nightside. These are shown averaged together and superimposed on the theoretical spectrum in Figure 7.5. The previously unobserved 0.85- and 0.90-micron windows are clearly present, and have about the predicted radiance levels.[4]

The quality of the data was not good enough to say very much about the surface mineralogy, but it did show that the windows might be employed by a more sensitive instrument on a future mission for that purpose. On the dayside, VIMS recorded several atmospheric absorption bands in the strong signal obtained with reflected sunlight as the source. Among those clearly identified were carbon dioxide and water vapour as expected, and more interestingly some very subtle features that earlier Earth-based observers had claimed were due to traces of mercury sulphide (HgS) and hydrated iron chloride ($FeCl_2 \cdot H_2O$).

Both NIMS and VIMS went on to the outer Solar System and successfully fulfilled their missions at Jupiter and Saturn respectively. At the time of writing, VIMS is still operating and expected to continue until 2017. NIMS died, with the rest of *Galileo*, in a fiery entry into Jupiter's atmosphere on 21 September 2003 at the end of a seven-year mission in orbit.

---

[4] The 'radiance' is a measure of the brightness of the emission from the planet at each wavelength, which on the nightside is the glow of heat radiation from the surface and lower atmosphere.

Chapter 8
Europe and Japan join in: *Venus Express* and *Akatsuki*

The first European mission to Venus had its origins in a Russian mission to Mars.

In 1996, a huge payload lifted off from Baikonur Cosmodrome, bound for the red planet, consisting of an orbiter, two 'surface stations' destined for a soft landing, and two penetrators which would impact the surface at high speed in order to burrow to a considerable depth and make measurements of the subsurface material. On board the orbiter was a number of scientific experiments built by European, mainly French, scientists who were collaborating with their Russian colleagues to explore the surface and atmosphere of Mars.

The launch on 14 November 1996 went well at first, but then everything was lost. The third stage of the booster was supposed to burn twice, once to achieve a temporary orbit around Earth and then again to align the trajectory towards Mars. The second burn failed and the scientific payload, over 6 tons of it, plummeted back to Earth.

The political situation in the eastern bloc was, by then, such that there was no prospect of rebuilding the mission and trying again. Instead, the Europeans looked into a project of their own which would use a smaller satellite to carry duplicates of the instruments they had built for the Russian mission. This could get to Mars in 2003; *Mars Express* was born.

The Mars mission was ultimately so successful that the European Space Agency (ESA) contemplated a follow-on, and in 2001 they issued an 'Announcement of Opportunity' to the scientific community asking for ideas as to what form it should take. The rules were that the basic spacecraft and launcher had to be the same as for *Mars Express*, but the instruments carried and the destination could be different. In the event, most of the responses proposed some kind of follow-up Mars mission, with a new set of instruments designed to build on the results from the first flight. However, the winning proposal advocated using the same spacecraft and some of the same instruments as had flown to Mars, but this time to go to Venus.

The *Venus Express* mission proposal

The scientific focus of the *Venus Express* proposal was on the planet's cloudy atmosphere, and its inhospitable and extreme surface climate. If the spacecraft was placed in a very elliptical polar orbit, remote-sensing instruments could cover the atmosphere and the surface from various distances, and travel through different parts of the magnetosphere to measure the field strengths and numbers and energies of particles. This made sure it would have wide appeal to a range of interested scientists from various disciplines. Much was also made of the fact that *Venus Express* would be the first mission to make use of the near-infrared transparency windows from orbit to become the first to carry out systematic remote-sensing observations of the Venusian atmosphere below the clouds.

The formal list of goals was given as:

- Study the atmospheric temperature fields above, in and below the clouds:
  - make observations of global temperature contrasts and the general circulation of the atmosphere;
  - investigate their coupling with cloud density and minor constituent abundance variations.
- Study the zonal, meridional and vertical motions at various levels:
  - employ studies of the movements and morphology of the features at different near-infrared window wavelengths, with high spatial resolution and long-term coverage;
  - determine the poleward extent of the low-latitude Hadley cell, with important implications for models of the general circulation and processes affecting the maintenance of the cloud-level super-rotation.
- Seek evidence for active volcanism and its extent:
  - obtain a better quantification of the volcanic gas inventory in the atmosphere;
  - use high-resolution near-infrared spectroscopy to obtain information on the species known or suspected to be variable in space and/or time, for example carbon monoxide, sulphur dioxide and water vapour;
  - make a detailed study of these species, along with temperature mapping of the surface and lower troposphere, to look for volcanic activity.
- Seek an improved knowledge of vertical cloud structure, microphysics and variability:
  - obtain detailed data on variations in cloud profile and opacity, including time-resolved, long-term data that allows the study of clouds as dynamical tracers.
- Produce updated inventories of minor constituent abundances:
  - make measurements of the variability in the distributions of CO, H_2O and sulphur-bearing gases;
  - carry out model studies of their role in cloud formation and the greenhouse effect.
- Carry out mapping of the general circulation:
  - investigate dynamical phenomena such as the polar vortices and deep atmosphere 'weather'.
- Obtain improved estimates of atmospheric loss rates for O, C, H and D:
  - quantify the main exospheric escape processes and model the long-term effects on climate change.
- Detect any interannual and interhemispheric asymmetries and trends in all of the above.

The proposers were at pains to point out that, despite the fact that *Venus Express* would be the twenty-eighth spacecraft to arrive at Venus since *Mariner 2* in 1962, the exploratory and innovative aspects of the mission meant there was still considerable scope for new discoveries.

Together, the expected and serendipitous findings would be used as a basis for:

- producing improved greenhouse models of the energy balance in the lower atmosphere;
- validating and improving general circulation models of the atmosphere, with improved treatment of the zonal super-rotation, the meridional Hadley circulation and the polar vortices;
- generating new climate evolution models using simple physics constrained by measurements; and
- comparative studies in all three areas with the other terrestrial planets including Earth.

They admitted that *Venus Express* could not address, let alone resolve, every one of the key questions about Venus that had accumulated as a result of exploration by the *Venera*, *Vega*, *Pioneer* and *Magellan* missions. The proposers candidly stated that knowledge gaps were likely to remain even after the new mission had been carried out, especially in the study of:

- aspects of atmospheric evolution requiring accurate measurements of noble gas isotopic ratios;
- surface-atmosphere interactions, requiring trace constituent abundance measurements near the surface;
- cloud chemistry, requiring direct sampling in the clouds;
- surface geology, geochemistry and interior structure, for which long-lived stations on the surface and sample return are the optimum way forward.

The scientific payload

If *Venus Express* was to address all these bullets and establish a new picture of the climate on Venus it would need a powerful suite of spectral and imaging instruments for the atmosphere and particle and field measuring sensors to probe the plasma environment. They had to be capable of carrying out a versatile and programmable set of orbital observations, including global monitoring and close-up imaging of atmospheric phenomena, solar, stellar and Earth radio occultation to study the structure and composition of the atmosphere, and *in situ* measurements of neutral atoms, plasma and the magnetic field. Everything had to be made available relatively quickly, and at a reasonable cost, to meet the Space Agency's criteria for going ahead with the project.[1]

The spacecraft did reuse the *Mars Express* bus, with a number of modifications for the different thermal environment at Venus (Plate 13). It also had a smaller communications dish, which could keep similar data rates to those from Mars because of the smaller distances involved to Venus. The final payload selection included four remote-sensing experiments for measurements of atmospheric properties, motions and surface mapping, and a magnetospheric package to seek, among other goals, new findings about the loss rates of atmospheric gases to the solar wind. Plasma probes and a radio occultation

---

[1] *Venus Express* was in fact cancelled for a short time, when it seemed it was going to add to the schedule and budget problems that ESA had across the board at the time. However, there was a fortuitous change of management at the Agency that led to restitution and reinstatement.

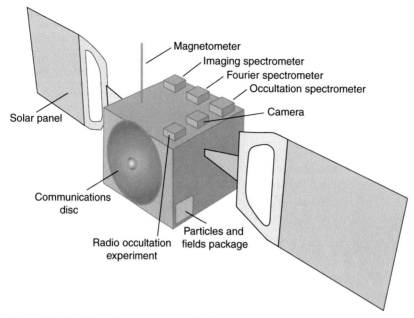

Figure 8.1 The *Venus Express* spacecraft, showing the scientific experiments it carries. The main body of the spacecraft is approximately a 1-metre cube (see also Plate 12, where it is compared with *Pioneer Venus*).

package to measure temperature profiles completed the payload. Of the seven scientific instruments, five were inherited from *Mars Express* and the *Rosetta* comet mission, and two were new. Here is a brief description of each one. Figure 8.1 shows how they are mounted on the spacecraft.

The Visible-Infrared Thermal Imaging Spectrometer (VIRTIS) is actually a compound instrument, with one spectrometer that maps Venus with moderate spectral resolution but good imaging coverage, and a second one that provides high-resolution spectra at a single location. Both worked in the ultraviolet, visible and near infrared out to 5 microns. The VIRTIS field of view ranges from a few hundred metres at closest approach (pericentre), to 15 kilometres at apocentre. The instrument used miniature mechanical refrigerators to cool the detectors to achieve the high sensitivity necessary for measuring the weak night-side emissions in the transparency windows. The mapping capability could track cloud features and temperature and compositional variations repeatedly, to make movies and get information on the atmospheric dynamics.

The Planetary Fourier Spectrometer (PFS) is the largest instrument in the payload and the only one not to work as planned. It was meant to cover the spectral range all the way out to 45 microns, with high spectral resolution and a field of view of about 10 kilometres at pericentre. The main scientific objectives were to study temperature, composition and aerosol properties in the middle and lower atmospheres, but none of this was achieved because of a jammed bearing in the scan mirror at the front of the optical chain. This mirror is a relatively simple device, compared with the rest of the instrument, used to point the direction of view at the selected location on the planet, and occasionally inside the housing at a calibration target, or at cold space to get a zero signal for reference.

The instrument produced beautiful spectra, but because it was stuck in the stowed position these were only of the calibration target it had been intended to view from time to time. If the mirror could be moved just once, to force it into a position where it viewed the planet, it could have been left there and the calibration sacrificed. Massive efforts were made to use special command sequences to free the mirror, even in desperation hammering it thousands of times in rapid succession, and turning the spacecraft to face the Sun so that the bearing heated up and expanded. Nothing worked; apparently the special spaceworthy lubricant in the bearing had been badly chosen and worked instead as glue after a few months in space. All that was left for the team was to reprogram the other spectral imaging experiments to try to plug the gap and partially recover the PFS scientific objectives.

Spectroscopy for Investigation of Characteristics of the Atmosphere of Venus/Solar Occultation InfraRed (SPICAV/SOIR) is another composite instrument package. It combines three ultraviolet and near-infrared spectrometers to study the vertical structure and composition of the mesosphere and lower thermosphere using solar and stellar occultation (Figure 8.2). This technique offers particularly high sensitivity to the abundance of

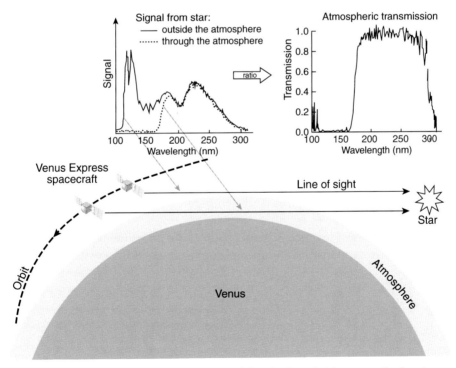

Figure 8.2 Stellar occultation measurements work by viewing a bright star, or the Sun, in space and then as it passes behind the limb of the planet. The gradually increasing absorption by spectral lines of various species in the atmosphere can be followed all the way down to the point where the signal vanishes.

minor species, including relatively scarce ones like several of the isotopes of the commoner gases. The ratio of deuterium to hydrogen in the isotopologues of water (i.e. H_2O and HDO) could be measured accurately in the search for new insight about the evolution of an ocean on early Venus. Because of the lower detection threshold compared with conventional measurements, it can also search for trace gases that were too scarce to have been picked up before. Among those targeted were certain hydrocarbons, including methane and ethane (CH_4 and C_2H_2), nitrogen oxides (NO, N_2O) and chlorine-bearing compounds (CH_3Cl, ClO_2).

The Venus Monitoring Camera (VMC) is a wide-angle camera for observations of the atmosphere and the surface through four filters in the ultraviolet, visible and near infrared. The spatial resolution ranges from 200 metres at pericentre to 50 kilometres at apocentre. The main goal is to investigate the cloud morphology and atmospheric dynamics by tracking the cloud features at various depths.

The Analyser of Space Plasmas and Energetic Atoms (ASPERA) focuses on the analysis of the plasma environment of Venus and the interaction of the solar wind with the atmosphere. It comprises four sensors: two detectors of energetic neutral atoms, plus electron and ion spectrometers, which together can measure the composition and fluxes of neutrals, ions and electrons to address how the interplanetary plasma and electromagnetic fields affect the Venus atmosphere and identify the main escape processes.

The magnetometer (MAG) has two fluxgate sensors to measure the magnitude and direction of the magnetic field in the magnetosheath, magnetic barrier, ionosphere and magnetotail,[2] with high sensitivity and temporal resolution and to characterise the boundaries between plasma regions. MAG can also search for lightning on Venus by measuring the strength of electromagnetic waves associated with atmospheric electrical discharges.

The Venus Express Radio Science Experiment (VeRa) uses signals emitted by the spacecraft radio system in the X- and S-bands to sound the structure of the neutral atmosphere and ionosphere with a vertical resolution of a few hundred metres.[3] An ultra-stable oscillator is carried on board to provide a reference frequency for the signal from the spacecraft transponder, which changes as it passes behind the planet in response to changes in atmospheric temperature, pressure and composition at the tangent point.[4]

In addition to the teams associated with these experiments, the project includes a number of interdisciplinary scientists and supporting investigators, who bring additional

---

[2] A 'fluxgate' magnetometer uses an iron core to pick up the presence of a background magnetic field, which is then detected by current-carrying coils wrapped around the iron. This type of device was invented before World War II, when a version was successfully employed to detect submarines.

[3] X-band and S-band are radio frequencies often used for communications with spacecraft. X-band corresponds to a wavelength of 3.5 centimetres and S-band to 13 centimetres.

[4] The information acquired on composition includes the abundance of sulphuric acid vapour below the clouds, since H_2SO_4 absorption affects the transmission of the radio signal.

strengths in several categories, such as atmospheric radiative transfer calculations and energy balance models, inversion of spectroscopic and radiometric data to obtain temperature and species profiles and cloud parameters, and dynamical modelling of the Venusian atmosphere using general circulation models like those used for the Earth and other planets.

A long-term goal for the whole team is to represent the climate process on Venus in a time-dependent model that will incorporate the results from the *Venus Express* investigations, leading to improved theories about the origin of the present state of Venus's climate, and informed speculations about its possible future evolution. A better understanding, not only of conditions on our planetary neighbour, but also of why they appear to diverge so much from conditions on the Earth, would be the ultimate goal.

The flight to Venus

Venus Express was launched on 9 November 2005 by a Russian *Soyuz-Fregat* rocket from the Baikonur Cosmodrome in Kazakhstan. This was a commercial arrangement, chosen by ESA partly because the Russians are good at getting payloads to Venus, but mainly because it was much cheaper than using a European or American rocket. All went perfectly, and on 11 April 2006 it reached the planet, there to be manoeuvred over the next several weeks into in a polar orbit at a height above the surface of Venus that ranged from just 250 to more than 66,000 kilometres (Figure 8.3).

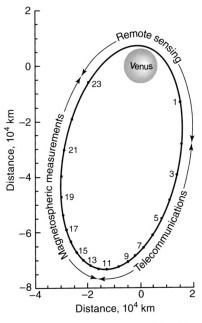

Figure 8.3 This scale diagram of the *Venus Express* orbit shows why most of the images are of the south pole and southern hemisphere – the passage over the north pole is very fast and correspondingly brief. The numbers on the orbit are hours from the time of the north polar passage.

This highly elliptical orbit meant that the remote-sensing instruments could collect data very close to the atmosphere when over the north polar regions, and from a distance of about ten times the radius of the planet when over the south pole. The thinking was to make use of the fact that, without seasons, the hemispheres of Venus should be expected to be near-identical mirror images of each other (there is some evidence accumulating that this is not quite true, in fact, but we don't know why) so the single spacecraft could combine global context and detailed, close-up views.

After about 50 days of checking that everything was functioning normally, the scientific mission of *Venus Express* began on 4 June 2006. This was planned to extend until 2 October 2007, corresponding to a lifetime of slightly more than two Venus sidereal days of 243 Earth days each, and rather more than two Venus years of 225 days each. The operators have to activate the small jets on the spacecraft regularly to keep the orbit from drifting away from the chosen values and decaying to the point where it would burn up in the atmosphere.

As well as good coverage, it was important to get high repeatability of measurements of dynamical phenomena, sufficient to make low-resolution 'movies' in which features can be tracked and their speeds and evolutionary characteristics identified. Time also had to be set aside for telecommunications to the Earth. The phase of the 24-hour orbit was maintained so that the ground station at Cebreros in Spain was always visible from the satellite for a fixed time each day for the downloading of data acquired during the previous orbit. Arranging this fixed time to correspond to the working day meant that expensive overtime costs could be minimised – financial considerations are ever present even on fully funded, operational projects.

During every orbit, the spacecraft and payload operate according to one of a series of predetermined 'science cases', each of which specifies the settings and data rate of each instrument as a function of time. The main purpose of this is to share out the limited amount of total data that can be stored and relayed to Earth. Generally, this meant that the remote-sensing instruments must choose between periapsis, off-periapsis and apoapsis viewing campaigns on any given orbit. Each of these has its own particular advantage: for about an hour and a half near periapsis, the altitude of the satellite is less than 10,000 kilometres, permitting high spatial resolution spectroscopic and imaging observations of the northern high latitudes. During off-periapsis observations, the spacecraft points to nadir or slightly off-nadir for eight hours to obtain a global view of the southern hemisphere, enabling spectral imaging of the motions of mid-latitude cloud features for studies of atmospheric dynamics. Near apoapsis, a further mode emphasises studies of the atmosphere in the south polar region, for example using VIRTIS to obtain movies of the vast polar vortex. Special science cases are sometimes provided to get the right geometry for things such as stellar, solar and radio occultations. Plasma and magnetic field measurements are obtained continuously on all orbits, to ensure maximum four-dimensional coverage of both the near-planet environment and the solar wind region.

Venus Express achievements

In more than 8 years of operation, *Venus Express* achieved extensive coverage of the atmosphere and surface of Venus, and achieved most of its objectives despite the loss of the Planetary Fourier Spectrometer. In reporting and justifying the (actually very modest, as space projects go) cost of the mission to its European political pay-masters, the science team (Plate 14) put together a list of key achievements, as follows.

(1) **The first global views of the double-eyed vortex at Venus's south pole.**

As we saw in Chapter 3, the *Pioneer Venus* orbiter infrared radiometer discovered the 'dipole' structure in the eye of the giant vortex over the north pole, and because it seemed to be always there, although its shape changed, it was expected that something similar would exist over the south pole as well. The *Pioneer* orbit was not set up to look in the south, while with *Venus Express*, the opposite was true (i.e. it could easily view the south but not the north pole) and the net result was spectacular confirmation of a similar vortex over both poles.

Not only that, but the increased sophistication of the modern instruments on *Venus Express* meant much more detail was obtained (Figure 8.4), and a much clearer idea of what produces the 'dipole' is emerging as a result (see Chapter 13).

(2) **The first detailed views of atmospheric structures such as clouds, waves and convection cells, in different regions**

This claim is based on pictures like that in Figure 8.5, which reveals fascinating details near the subsolar point, meaning the Sun is directly overhead. We would expect the heating to be greater here than anywhere else on the planet, and also to penetrate to greater depths, since the rays take the shortest path – vertically – through the cloud layers. The result is a patch of atmosphere that appears to be full of small convection cells, rather like a pan of boiling water. There is also a higher concentration of the dark ultraviolet absorbing material, suggesting that the boiling is bringing it up from below. The disturbance is shifted towards the afternoon by a couple of hours, corresponding to the sort of time lag we would expect if the maximum solar heating is 10 kilometres or more deeper than the level we are observing, near where the cloud is thickest.

The rapid zonal winds do the rest, moving and tilting the convection cells and at the same time responding to the obstruction presented by the subsolar disturbance by streaming around it and generating a huge global-scale wave motion. This is the origin of the massive 'sideways-Y' feature that was a conspicuous feature of the first ground-based ultraviolet observations, more than half a century ago.

Figure 8.4 A spectacular early result from *Venus Express* was the first detailed images of the south polar vortex. On the day this view was obtained the 'dipolar' nature of the vortex eye is subdued, but it can still be seen to be elongated into an 'S' shape. The picture is about 1,000 miles across.

(3) Detailed maps of wind fields and temperatures, yielding three-dimensional data about the structure and the dynamics of the atmosphere[5]

Wind 'fields' – that is, mapped over the planet, as opposed to the kinds of single profiles we get from tracking descending entry probes – are difficult to measure except at the

---

[5] This is actually a combination of three achievements from the original release, which was presented as a 'top ten'. It would be fairly easy to make the number up again with some additions – my choices would be the clarification of the reason for Venus's climatic state, and the new insights into the role of volcanism. Although it is not science per se, the mission also deserves applause for saving the scientific study of Venus as a discipline from extinction, with no prior mission for nearly twenty years and still none on the horizon in the future.

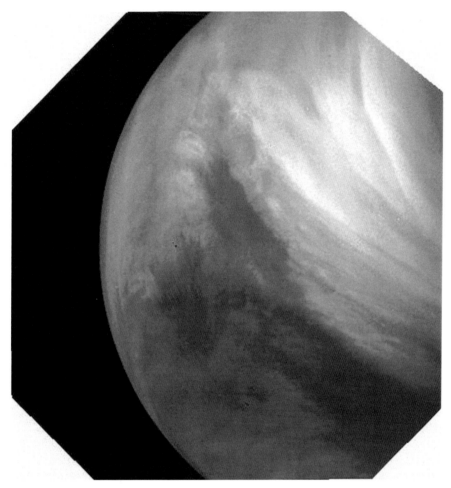

Figure 8.5 Detailed cloud structure seen in an ultraviolet image from the *Venus Express* camera. Local noon (Sun overhead) is near the centre of the dark region at left, the south pole is off bottom right.

cloud-top level where the dark features can be tracked in movies or simpler series of images. *Venus Express* had the advantage, because it was the first orbiter to exploit the recently discovered near-infrared windows, to look at more than one level by tracking different layers of clouds.

All sorts of interesting conclusions could be drawn from this, perhaps the most important being the finding that the winds, like the clouds, have two distinct regimes, one at low latitudes and the other polewards of about 50 degrees of latitude. Either side of the equator, the winds depend on height (consistent with the entry probe wind profiles) but not on latitude, but in the giant vortices they become the same at all heights, falling off towards the pole (Figure 8.6). The sharp transition between the two regimes, and the apparent symmetry between north and south, are quite remarkable.

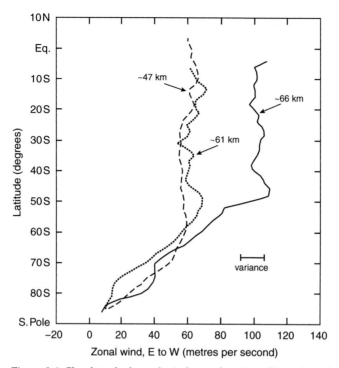

Figure 8.6 Cloud-tracked zonal winds as a function of latitude in the southern hemisphere, as inferred from *Venus Express* near-infrared images. The three curves correspond roughly to the winds in the upper, middle and lower cloud layers at the approximate heights above the surface shown.

(4) The first large-area temperature maps of the searing surface of the southern hemisphere.

These maps show a lot of detail, and contrasts of several tens of degrees (Figure 8.7), but most of this is associated with the highs and lows of the surface terrain following the vertical profile of atmospheric temperature. In some places, particularly on volcanoes (see Plate 10), the temperature does not match that expected from the height of the feature; the anomaly forms a well-defined shape, apparently a lava cap.

Unfortunately, we cannot simply deduce from this that the lava is hot, and therefore that it was recently discharged from the volcano. All of the values measured are a mixture of temperature and emissivity effects, and it would only be possible to say with certainty that the lava was actually hotter than its surroundings if the calculated emissivity that fit the data was more than the theoretical maximum value of 100 per cent. Such a case has not been found yet, so the safer inference is that the composition of the lava cap is different from its surroundings, with a different emissivity. This, too, is interesting: the higher-emissivity material is probably less weathered and therefore younger. It might be hotter as well. Towards the end of the

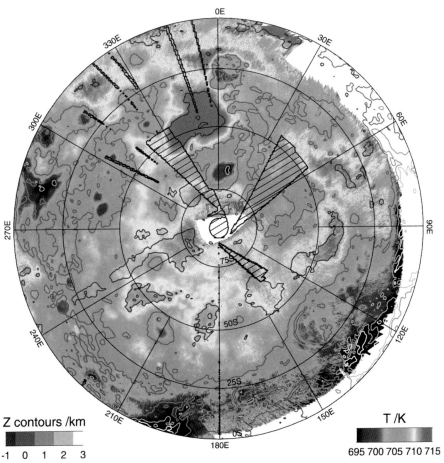

Figure 8.7 Surface temperature anomalies (shaded) in the southern hemisphere show a strong correlation with the height of the terrain (shown by the contour lines).

mission evidence began to emerge that some of the features were changing with time, and even vanishing altogether, further supporting the idea that they are the results of volcanic eruptions.

(5)   **The most complete data set to date of the chemical species in the atmosphere.**

Since it did not have mass spectrometers on entry probes, like *Pioneer* and *Venera*, and the high-resolution Planetary Fourier Spectrometer instrument failed without obtaining any data, this claim is not as strong as it might have been. However, the occultation spectrometer, SPICAV, obtained spectacular new data in the middle and upper atmosphere, including new and much improved deuterium-to-hydrogen ratio profiles, of key importance for understanding the history of water on Venus, and amazing observations of enormous variations in the amounts of sulphur-bearing gases such as SO_2, presumably originating in volcanic activity on the surface nearly 100 kilometres below.

(6) **The water escape rate from the atmosphere in relationship to the bombardment by the solar wind.**

Atmosphere is blasted away from Venus under the effect of the solar wind, a stream of high-energy particles from the Sun (Figure 8.8). The interaction is very different from Earth, because of the absence of a planetary magnetic field on Venus to deflect the charged particles and focus them into the polar regions. Whether this leads to more or less atmospheric erosion is a subject of debate at present: the old idea that Earth's field was a shield against

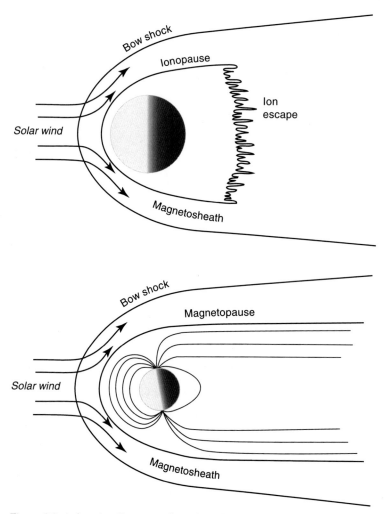

Figure 8.8 A drawing illustrating how the solar wind blows molecules off the top of Venus's atmosphere, above, without having first to interact with a planetary magnetic field as it does at Earth, below.

solar wind erosion of the atmosphere has been challenged since *Venus Express* began to map out the magnetosphere and its complexities. It might instead trap and funnel the energetic particles down into the atmosphere. The jury is still out on this.

In either case, Venus is losing many tons of gas from its atmosphere every second. The spacecraft measured the amount and composition as it passed through the tail of atoms, molecules and ions that points away from the Sun, and confirmed that most of it is the lightest species, hydrogen and helium (Plate 15). What was a major surprise was that the instrument teams reported a lot of the relatively heavy oxygen as well, and suggested that H and O might be in a ratio close to two to one, from which it follows that the source is water that has been dissociated in the upper atmosphere. The problem is that the solar wind-magnetospheric interaction is so complicated in space and time that a single space-craft, even one that operates for several years like *Venus Express*, cannot produce a completely reliable number for the rate of loss of all species from the whole planet. Most of the loss of water may occur in relatively isolated, but very powerful, bursts of particle radiation from the Sun that are difficult to record.

(7)   **The most detailed view of the 'oxygen airglow' and 'carbon dioxide fluorescence' of Venus, which make the planet glow like a space lantern.**

These are faint emissions from the upper atmosphere (altitudes above 100 kilometres) that occur after the Sun dissociates molecules or excites atmospheric species into long-lived states. The fragments can recombine and produce species (such as molecular oxygen, O_2, and hydroxyl, OH), again in excited states that decay slowly, emitting ultraviolet, visible and infrared photons.

The excitation occurs on the dayside, but the prevailing winds at those high levels carry them to the nightside, so that most of the decay, and emission, takes place on the nightside near local midnight. In the near-infrared a cloud of oxygen has been observed near local midnight on the nightside of Venus. The gas is glowing because it emits photons at wavelengths close to 1.27 microns when two atomic oxygen atoms, produced on the dayside by the dissociation of carbon dioxide, combine to produce the familiar molecular form of oxygen, O_2.

Airglow is a well-studied phenomenon on Earth, but much less so on Venus. *Venus Express* was able to identify several emitting species that had not been seen on Venus before, including hydroxyl (OH) and nitric oxide (NO), and track the emissions as tracers of the dynamics of the upper atmosphere (Chapter 13).

(8)   **The first unambiguous detection of lightning in the atmosphere.**

This is perhaps the most controversial of the claims, and the latest in a long history of measurements and speculation (Figure 8.9) that are still not completely conclusive. The lightning experts on the *Venus Express* team, like many of their counterparts in *Pioneer* and *Venera* days (some of them are the same people), were confident that the 'whistlers' and low-frequency radio bursts that the spacecraft detected could only have come from light-ning at something similar to the Earth's level of activity. It has also been pointed out that the

Venus Lightning

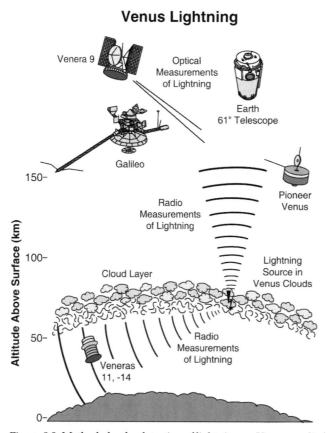

Figure 8.9 Methods for the detection of lightning on Venus applied by the various spacecraft, shown prior to the current studies by *Venus Express*.

detection of nitric oxide in the lower atmosphere is also evidence for lightning, because this is the only way that an intense enough energy burst can be found to break the strong bonds in molecular nitrogen to produce the oxide, as happens on Earth.

The old objection that prevailed for a long time and led to the belief that it was not even worth looking for lightning on Venus, no longer applies. The idea was that the Venusian clouds were too tenuous and too quiescent for lightning generation, according to the theoretical understanding of the processes of charge exchange and accumulation required. This argument prevailed until it was laid to rest by the discovery in the near-infrared windows of dense clouds and vigorous cumulus dynamics in the deepest layer. The lightning would have to be cloud-to-cloud, rather than cloud-to-ground, because the dense cloud clusters are typically 50 kilometres above the surface, much higher than the equivalents on Earth. This is not a problem however, as cloud-to-cloud is the commonest form of lightning even in the terrestrial situation where the surface is much closer to the source of the strike.

What is harder to explain is why there have been so few visual detections of lightning flashes on the nightside of Venus. Earth and Jupiter show abundant and spectacular displays of lightning to the cameras on orbiting spacecraft. On Venus, diligent attempts by the teams operating the *Venera* orbiters, the *VEGA* balloons, the *Pioneer Venus* star sensors and the camera on *Venus Express* have produced just one observation. That came from *Venera 9*, and has been disputed.

Ground-based observers likewise have managed just one positive report despite multiple attempts over a long period of time. Perhaps something about the conditions on Venus, or limitations of the observations themselves, has made lightning hard to spot, but until there is a breakthrough with multiple, confirmed sightings the jury will still be out, and new and improved detection attempts will be planned for future missions.

Venus Express lives on

Fortunately, since there is no successor in sight, *Venus Express* is proving to be a very long-lived mission. At the time of writing, the spacecraft and most of its payload are still operating, after no less than four extensions to the funding for mission operations since their successful arrival at Venus in 2006 (Plate 16).

Recently the risk has been taken of allowing the spacecraft to dip into the atmosphere during the lowest part of its orbit, down to an altitude of about 165 kilometres. The atmospheric drag is measured by the torque on the reaction wheels when the solar panels are set at right angles to each other, and the density profile is then worked out. As well as checking models of temperature and composition, such studies have an engineering application, since future missions plan to use the drag to achieve orbit or landings (the so-called 'aerobraking' technique).

Venus Express is scheduled to be decommissioned soon, no later than the end of 2015, and before that if the fuel used for maintaining the orbit is found to be running out. The actual amount remaining is surprisingly hard to measure, because there is no fuel gauge. The controllers have to try to work out how much has already been used, by adding up the sum of each firing. They can also resort to tricks such as rolling the spacecraft and estimating its moment of inertia, and from the distribution of mass the amount due to the fuel tank and its remaining contents. The two methods do not agree very well, but both suggested in late 2013 that the supply was getting low, and unlikely to last more than another year. The team then has to decide whether to make a planned end to the mission, perhaps including a suicidal final descent into the deeper reaches of the atmosphere to measure its density and temperature from the drag encountered, or to soldier on in orbit until the tanks run dry. The spacecraft will then drift, untracked and excommunicated, until it crashes into Venus and burns up in the atmosphere a few years later.

Japan's climate orbiter, *Akatsuki*

The Japan Aerospace Exploration Agency (JAXA) also had plans for remote-sensing investigations of Venus. Their 'Planet-C' spacecraft was launched on 20 May 2010, and its name changed from its development title of *Venus Climate Orbiter* to the more romantic *Akatsuki*, which means dawn or daybreak (Figure 8.10).

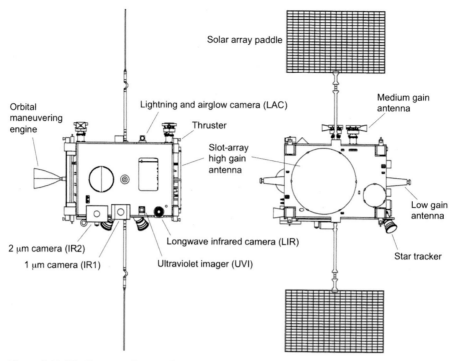

Figure 8.10 The Japanese *Venus Climate Orbiter* spacecraft, renamed *Akatsuki* after launch.

Like the European mission, this was to focus on the Venusian atmosphere, but with a stronger meteorological flavour. To carry out long-term monitoring of the dynamics of the atmosphere from orbit it has a suite of five sensors (the Japanese call them 'cameras', although they are rather more than that) exploiting different spectral ranges and windows, from the ultraviolet to the mid-infrared, to get access to different levels in the atmosphere (Figure 8.11). These would observe the planet from an equatorial orbit, and the period of the orbit was chosen to be a close match to the four-day circulation of the atmosphere.

The idea was to follow weather systems in the atmosphere and study continuously how they evolved, rather than getting disconnected glimpses as a polar orbiter like *Venus Express* does. The price that is paid is the loss of high-latitude coverage, and coverage of the fascinating polar vortices. The Lightning and Airglow Camera was designed to look for discharges at visible wavelengths, the Long Wave Radiometer would study the structure of high-altitude clouds in the thermal infrared near 10 microns wavelength, and the UltraViolet Imager would monitor the distribution of sulphur dioxide. Two near-infrared radiometers covered the near-infared window wavelengths to map the surface and sound the lower reaches of the atmosphere.

The spacecraft launched successfully on an H-IIA[6] rocket and had a good flight to Venus, firing its orbital insertion motor on 10 December 2010 just before midnight GMT to

---

[6] The H-IIA is built for JAXA by Mitsubishi and has a payload capacity to geostationary transfer orbit about the same as that of the American Atlas-V and the European Ariane-V.

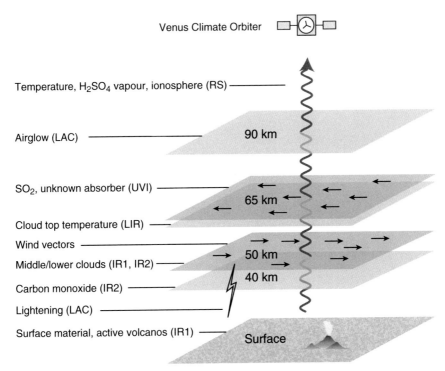

Figure 8.11 A schematic representation of the *Akatsuki* objectives.

slow down and achieve capture by Venus's gravitational field. During the planned 12-minute burn, the spacecraft passed behind the planet and communication with the Earth was lost.

This of course was expected, but when it emerged the spacecraft was soon seen to be on the wrong trajectory and rapidly leaving Venus behind. The problem, it later emerged, appeared to be that the rocket motor had a fault and was delivering only about 10 per cent of the thrust it was designed to provide.

Thus *Akatsuki* ended up in orbit around the Sun instead. All may not be lost, because the trajectory can be trimmed so that the spacecraft will meet Venus again, after several trips around the Sun, and a fresh attempt will be made to achieve Venus orbit in 2015.

Part II
The motivation to continue the quest

Having arrived at an up-to-date position on missions to Venus, we can now summarise what we know, and ponder what we still wish we knew, about this close-by and Earthlike world. To begin with, as we have seen, in some ways it is not much like the Earth at all, and we seek to find out why.

Since 1962, there have been no fewer than forty-four spacecraft launched towards Venus, to orbit, land, float or just fly past and make observations in transit (although not all of them were successful, see Appendix A). This programme, along with advanced Earth-based observations over a wide range of wavelengths, and some imaginative theory and comparative planetology studies (mostly treating Venus as an analogue of the Earth, recognising that Mars and Mercury are siblings too), has painted a fairly complete general picture of our mysterious neighbour at last.

In some ways, this vision is not too encouraging, at least for those who see planetary exploration as a search for Earthlike, habitable environments. Where Mars turned out to be a cold place with a thin atmosphere, Venus is now seen to be the opposite, with a surface environment far worse, for human survival, than the conditions found in a pressure cooker in any kitchen on Earth. In fact, if a tin of beans was placed on the surface of Venus (and if the tin were really made of tin, which they are not these days) the atmosphere would not only cook the beans but the tin would melt as well. Human expeditions are not therefore on the cards for Venus for quite some time, and the prospects for any kind of life there have faded (but not quite vanished) after centuries of raised expectations based on early twentieth-century predictions of tropical forests and warm, soda-water oceans.

This change in perception does not mean that we should now see Venus as dull; far from it. First of all, the processes giving rise to the ovenlike climate at the surface turn out to be remarkably similar, indeed in most respects identical, to the carbon dioxide-driven greenhouse phenomenon that is threatening to warm the Earth to uncomfortable and even dangerous levels for the inhabitants here. The struggle to understand and forecast global warming on Earth has been given a big boost by having another example close by and available for study, even, perhaps especially, if the portents are scary. Other phenomena on Venus also have earthly parallels – complex clouds and weather systems, high winds, polar vortices resembling that over the Antarctic – but there are puzzles too, for example the slow rotation that produces a very long day–night cycle, and the reasons for the missing Venusian magnetic field. These are things that, before we knew the truth, were reasonably expected to be more like Earth's.

The processes that shaped the surface of Venus, including the role, if any, of the movement of large-scale plates analogous to 'continental drift' on the Earth, remain mysterious. Volcanoes are everywhere and may still be actively maintaining the cloud layers and thick atmosphere, and hence the extreme climate of Venus. If so, the climate is probably subject to change, but nothing about this is certain, including the timescale and the ultimate state, which could conceivably be much more Earthlike. It is likely that evidence is preserved in the subsurface layers that will shed some light on the question of whether or not Venus once had oceans, and perhaps even life. We might be on the verge of a comprehensive understanding of the origin and maintenance of the zonal super-rotation, and its relationship to the complex polar vortex dynamics. The cloud structures seen in the near-infrared window images are witness to a zoo of meteorological activity in the deep lower atmosphere of Venus, which had previously not been glimpsed or expected.

Can we devise experiments that are capable of giving new insight into these and other key topics, at a reasonable cost and risk? Planetary science groups in universities, government research labs and space agencies around the world ask these questions often, and make plans that are constantly under revision. In the next part of the book we look at what they are thinking, topic by topic.

Chapter 9
Origin and evolution: the solid planet

Twin planets?

When considering the solid planets and not their climate at the surface (which is the subject of Chapter 10), Earth and Venus seem very similar and really are twin planets. But they are not identical twins. We have already seen that Venus is 5 per cent smaller in diameter and quite different in its orbital dynamics, with slow retrograde rotation and near-zero obliquity. The absence of a magnetic field is one of the few really clear indications that we have that the deep interiors are not altogether the same. The absence, or at least the difference in character, of plate tectonics on Venus suggests major differences in the solid crust, as do substantial differences in the geography of the two surfaces.

It is a crucial part of our outlook on the world, and not just as scientists, to try to understand to what extent Venus and Earth are the same and to what extent they differ, and why. To a very great degree, we still lack the experimental evidence to answer this, although some progress has been made. When considering the solid body, progress will continue to be slow because it is much harder to make surface and, especially, interior measurements of the kind we need. This is in contrast to the atmosphere, which is now quite well studied because it is relatively accessible. Unless there is a surprising increase in the amount of effort and money devoted to planetary exploration and prospecting, it will be a long time until we have data from deep-drilled cores and seismological measurements on Venus that are comparable to those that are responsible for so much of our knowledge about the earth below our feet.

In the meantime, theoretical models of solar system formation, and arguments based on what we know about the composition of the universe at large, along with the data we do have, help us to break down major mysteries into more focused questions and to begin to plan missions and projects to address them.

What is Venus made from?

Of course, we do not know with any certainty the relative abundances of the chemical elements in Venus as a whole, since we cannot sample the interior, and we have literally only scratched the surface. To a lesser extent the same is true for the Earth as well. But there are various ways in which we can infer these values and come to conclusions that we think are probably not too far from the truth, at least as far as the commonest substances are concerned. The starting point is the cosmic abundance of the elements: looking around the universe, and collecting and analysing meteorites that fall to Earth, we are able to infer the mix of hydrogen, helium, carbon, oxygen, sulphur, magnesium, iron and the rest that must have made up the cloud that became our planetary system.

Then we look at relevant parameters that we can measure. Venus is 320 kilometres smaller in radius than the Earth (about 5 per cent), and less massive by about 18 per cent. This makes the mean density of Venus 5.24 g/cm³, which is 5 per cent smaller than that of Earth. However, when considering how their bulk compositions compare, what counts is the uncompressed mass: even rock and iron increase in density under the enormous pressures found deep inside an Earth-sized planet, and of course the effect depends on the total mass and its internal distribution. With a few assumptions the uncompressed densities can be worked out, and lo, they are the same at 3.9 g/cm³. This, plus expectations based on how we think the planets of the Solar System formed and the fact that Venus obviously has a basaltic crust like the Earth, leads us to assume that the planets are pretty much the same inside in terms of composition and structure.

The outermost layer, the crust, is solid rock but only a few tens of kilometres thick on average, so it can be lifted, moved and cracked by convective plumes in the deep mantle of liquid rock that lies underneath (Figure 9.1). The images obtained by the radar-mapping satellites show that there are significant differences between the way the surface behaves on Venus and on Earth, probably mainly as a result of the relative thinness and dryness of the crust on Venus. The high temperature and limited amount of water vapour in the atmosphere, and especially the absence of liquid water oceans, mean the Venus crust has been baked into a hard and rigid state.

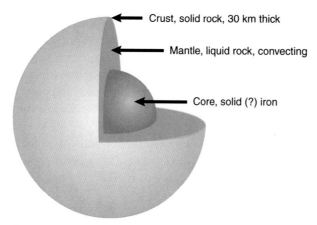

Crust, solid rock, 30 km thick

Mantle, liquid rock, convecting

Core, solid (?) iron

Figure 9.1 The probable interior structure of Venus, as inferred from its size and density, likely composition and theoretical models. The solid crust forms the top 30 kilometres, some of it exposed to the atmosphere. The rest of the rocky shell is the molten mantle, while the innermost 50 per cent or so of the diameter is the metallic core, made principally of iron and nickel.

The large plates that slide slowly around the surface of the Earth, lubricated by water, giving us continental drift, seem to be replaced on Venus by a single shell covering the whole planet. This is about 30 kilometres thick according to recent estimates, but somehow strong enough to support enormous volcanic mountains in various places. This can be seen in the gravitational signature measured by perturbations in the orbits of satellites, tracked from the Earth: these are large on Venus, showing that a huge mass is being supported above the average radius. On Earth, the mountains sag and depress the crust, and the orbital perturbations are smaller.

The extra lift on Venus must come from convection in the mantle, providing dynamical uplift in slow currents of liquid rock that rise from the interior, carrying heat away from the core. The largest constructs on the surface sit on top of the most vigorous of these plumes, and release some of the liquid rock and accompanying gases into the atmosphere as volcanoes.

Composition of the surface rock

In principle, we can learn about the composition of the crust by studying the rocks on the surface, either directly after landing or by remote sensing from orbit. Other clues come from analysing the effect of surface-atmosphere interactions on the composition of the atmosphere, but these have large uncertainties. The problem, of course, is that compared with Earth, Moon or Mars, the surface of Venus is seriously inaccessible to direct probing or sampling, while as long as we are restricted to doing it from orbit, remote sensing is hampered by the thick, cloudy atmosphere.

We do have data from the X-ray and gamma-ray instruments on the *Venera* and *Vega* landers, but these, while an extraordinary achievement, fall far short of what is needed for several reasons. First, the number of samples is very small – just seven locations altogether (shown on the map in Figure 2.7), all of them in the equatorial regions and all on the geologically young lava plains. It will be essential eventually to have surface measurements in some of the ancient, rocky uplands, but the early landers were, of course, more obsessed with getting down safely and less with sampling geochemical diversity, and volcanic plains cover around four-fifths of the surface. The landers had very limited targeting capabilities because they were effectively thrown at the planet, rather than steered, and during the descent the trajectory was further altered by strong side winds. Most of them were confined to low latitudes by the energetics of their trajectories from Earth to Venus,[1] which of necessity are in or near the ecliptic plane in which the planets orbit.

Even after the fact, the locations of the places at which the spacecraft touched down can be reconstructed only to within an error of about 300 kilometres. Since the *Magellan* radar images have a resolution better than a kilometre, this uncertainty is a major limitation in coordinating the two when trying to analyse what the samples represent. Future landers

---

[1] The *Pioneer Venus* project managed to get its small probes (each with a mass of 90 kilograms) to fairly high latitudes, including one at about 60 degrees north, but the larger probes all landed near the equator, see Figure 2.7.

on the highlands of Ishtar or Maxwell Montes, say, will require a powered approach to reach higher latitudes, and a means of navigating during the descent through the thick, windy atmosphere with GPS-type guidance as they do so. We have recently seen progress towards this on Mars with the 'sky crane' used to land the *Curiosity* rover in 2012, with plans for a network of communications and guidance satellites there in the future.

Once the historic Venus landers were on the surface, the difficult operating conditions meant that only fairly crude measurements could be made and that even these would have significant errors and uncertainties. The *Veneras* measured the ratio of potassium, thorium and uranium in the rocks at the landing site, and in some cases the abundances of most of the important elements that make up the commonest kinds of rocks on the Earth, including magnesium and aluminium but not the interesting trace metals like chromium and nickel. The results are consistent with various types of basalt, as expected, but even with all of the limitations of coverage and precision they showed that Venus is very diverse in its mineral composition. For example, some areas are very rich in the radioactive elements and others highly depleted; one sample indicated lava that had been modified in some way that some analysts suggested indicated fluid activity, possibly even the action of water (somehow).

Very limited progress has been made in obtaining surface composition information by remote sensing, for example with the infrared mapping spectrometers on *Venus Express*, which identified relatively young lava flows (Chapter 8). This, along with *Galileo* and *Cassini* (Chapter 7), showed what could be done from a floating platform, the lower the better to minimise the atmospheric hazes and absorptions. Differences in reflectivity and emissivity at radar wavelengths between regions, for example the 'snowcaps' on the highest mountains, are, like the lander data, intriguing rather than conclusive.

While it is amazing that we have anything at all, it remains the case that the definitive measurements of Venus's surface composition have yet to be made. Working out how to get this vital information is one of the key opportunities for experimenters on future missions.

Messengers from the interior: the noble gases

A frequent theme in proposals for new missions to Venus is the careful measurement of certain gases in Venus's atmosphere that are present in extremely small amounts. At the top of the list usually come the noble gases, helium, argon, neon, krypton and xenon. These are very chemically inert (hence 'noble') and so have generally not been altered by chemical combination, in contrast to the more common elements, because they were produced or trapped in the crust. In addition, they have a rich variety of isotopes – variants of slightly different mass – and the ratio between these is a valuable indicator of the history of the gas. That history generally includes outgassing from the interior, and so it has implications for the whole planet (Figure 9.2).

For instance, argon-40 is radiogenic (produced by the decay of radioactive potassium-40), neon-21 is nucleogenic (produced from neon-20 which absorbs a neutron emitted by radioactive elements such as uranium) and xenon-134 is fissionogenic (produced when

Species and atomic numbers	Primary Contribution to understanding the origin and evolution of a planet	Current Estimate	Uncertainty
Noble Gas Abundances		Parts per million	
Xenon 132	Relative contributions of sources and sinks of material, e.g. blowoff, comets and planetesimals	0.0019	200%
Krypton 84	Role of comets in providing atmosphere	0.0004	400%
Argon 36	Comets vs. planetesimals in planetary formation	31	33%
Neon 20	Common kinship of Earth and Venus	7	50%
Helium 4	Role of outgassing from interior	12	200%
Isotopic Ratios		**Ratio**	
Xenon 129/130	Estimate loss of atmosphere by blowoff	3	100%
Xenon 136/130	Test the uranium-xenon hypothesis	1	100%
Neon 40/36	Estimate outgassing into the atmosphere	1.1	12%
Argon 36/38	Role of large impacts	5.4	200%
Neon 21/22	Earth and Venus: twin planets?	< 0.067	200%
Neon 20/22	Role of hydrodynamic escape	11.8	6%
Helium 3/4	Role of the solar wind	< 0.0003	200%
Deuterated/ordinary water (HDO/H_2O)	Loss of water over time	0.019	32%
Nitrogen 15/14	Loss of total atmosphere over time	0.0037	22%
Sulphur 34/32	Current volcanic activity	0.04	100%

Figure 9.2 Noble gas abundances and isotopic ratios on Venus and some of their implications.

uranium itself decays).[2] It is immensely complicated to work out (xenon alone has eight stable isotopes, each with its own story to tell), but cosmochemists can use isotopic ratio measurements to deduce details of the formation, composition and outgassing history of the planet and compare them with Earth, Mars and meteorites.

Knowledge of the isotopic ratios in the solid materials of the crust is just as important as that of the gases in the atmosphere, but harder to obtain. Eventually, drilling to considerable depths and returning core samples to Earth for analysis will reveal much, but in the meantime the isotopic ratios in the atmosphere are being targeted as they are so much more accessible.

Even so, because the abundances are very small in some important cases, and the measurements need to be made with great precision to be most useful, it would be better to collect atmospheric samples and return them to Earth than to try to analyse them on Venus. Much larger and more sensitive devices are available in the laboratory than on a space probe. For example, a state-of-the art mass spectrometer fills a large room and weighs 100 times more than the largest spaceborne version. Also, of course, conditions are

---

[2] The numbers are the atomic masses, in units where the mass of hydrogen equals 1.

more benign and stable as well, since the laboratory is not hurtling through clouds of corrosive acid while trying to make delicate measurements.

The noble gases are sometimes called the 'rare' gases on Earth. They are also extremely scarce in Venus's atmosphere: there is less than 1 part per million for all six of the stable isotopes of krypton, for instance. The abundance of xenon on Venus is not known at all, but is probably less than this. Neon is about 5 parts per million. On Earth, neon has a higher proportion of its heavier isotopes in the atmosphere than it does in the crust, indicating fractionation (faster loss of the lighter isotopes) during the escape of neon from the top of the atmosphere in the past.

We have still to measure these ratios on Venus well enough to see if they tell a similar story. On Earth and Mars the sum of all of the isotopes of xenon adds up to less than one-hundredth of the amount of gas that we would expect to find based on cosmic abundances, as found in meteorites and elsewhere. On Venus, we have yet to find it at all until we can make more sensitive measurements. The reason has to be that a lot of xenon has escaped, but this is remarkable since, with an atomic mass of 129 for the commonest isotope, xenon is one of the heaviest gases in the atmosphere. If more than 99 per cent of it has been lost over time, there are obvious implications for neon (mass 20), nitrogen (28) and carbon dioxide (44) as well. The most probable explanation is that one or more giant impacts blew off heavy and light gases alike, followed by renewal of the atmosphere with gas richer in the lighter elements by emission from the interior.

Helium, at 12 parts per million, and argon, at 70 parts per million, are the most abundant and the best studied noble gases. The helium on the Earth seems to be made up of two distinct sources (three if the flux from the solar wind is large enough to be important), one from the deep interior, which is rich in primordial helium-3, and the other produced radiogenically and outgassed from the crust. The release processes on Venus are different (if there are no plate tectonics, for example) and will give a different mix; there may also be more solar wind helium in the atmosphere because of the lack of a magnetic field to divert the flow as it does on Earth.

Argon-40 is less abundant on Venus than on Earth, but there is a lot more of the non-radiogenic argon-36 on Venus. It is puzzling why that should be. The ratio of argon to neon seems to rule against the Sun as the source, which might have made sense given that Venus is closer. Instead, the isotopic ratios suggest to those who study these things that the excess neon on Venus came from volatile-rich meteorites. Perhaps Venus had a collision with a very large comet that altered the composition of the whole atmosphere. Questions like this, and many others, could be addressed much better were it not for the rudimentary state of the current measurements.

The abundant light elements like carbon, nitrogen and oxygen also have stable isotopes, although fewer than the noble gases. These too show great diversity between the planets, for reasons related to how they were formed (nucleosynthesis), their physical distribution in the solar nebula before the planets formed, and different loss processes (escape into space, for instance, or chemical combination with surface solids) during formation or afterwards. The case of deuterium (heavy hydrogen) is special because of its relationship to the history of water on Venus, and we return to this in Chapter 10.

Why no magnetic field?

Another big difference between the interiors of Venus and Earth is that whereas processes inside the Earth produce a strong magnetic field, Venus has none at all. Attempts to find internal magnetism at Venus have now reached down to the measurement limit where a field about a million times weaker than Earth's would have been detected had it been present. The difference between the planets is surprising given that the similar high mean densities of Venus and Earth must mean that Venus has a large metallic core like the Earth, while evolutionary models and the apparently high level of recent volcanic activity both suggest that this core is still in a partially molten state. Heat has to escape outwards, leading to convective motions in this molten core. Why is this not associated with an internal dynamo? It is often assumed that the reason must be something to do with the slow rotation of the solid body of Venus, although why that should be does not stand up to close scrutiny and the experts on magnetic fields mostly reject the idea.

Perhaps Venus did have a field once, but it is currently in the null state between reversals. Reversals are quite common in the terrestrial magnetic field record. Or perhaps, despite our expectations, the core has somehow cooled efficiently and largely solidified. Conversely, some theories allege that Venus is too hot inside to generate a field. Finally, perhaps the randomness in the way we now believe the planets assembled means that Venus's core does not have the composition we expect from our knowledge of the cosmic abundances of the elements and the mean density of the planet, and is not magnetogenic.

More likely, the slightly smaller mass of Venus may be responsible for a significant difference from Earth. This could be the case if the compression at the centre is below the critical point for core formation. At the same time, if the interior of Venus lacks the cooling effect of efficient plate tectonics near the surface, it may be hot but still have only subdued convection currents in the core, so no measurable field is produced. If, in turn, the suppression of plate tectonics is a manifestation of the high temperature and extreme dryness of the crust, it may be the loss of its oceans that deprived Venus of its magnetic field.

In sorting through all of these possibilities, a key question to ask is whether Venus ever had a field, and if so when it was lost. As expected for Mars, the return of drill samples from the surface and interior of Venus may eventually shed light on this puzzle. There is a crucial difference, however. Unlike Earth and Mars, Venus might not have maintained any remnant crustal magnetism because the temperatures in the crust are above the Curie point for most of the common magnetic minerals.[3] We may never know the answer.

Venus's surface features: Earthlike but different

Following the breakthrough achieved by Earth-based radar astronomers, the global radar coverage provided by the *Pioneer Venus*, *Venera* and especially the *Magellan* spacecraft has

---

[3] The Curie point is the temperature at which the thermal agitation of the lattice of the magnetic material scrambles the alignments of the molecular dipoles that give the sample a macroscopic field. The dipoles do not realign upon cooling unless they are again subject to an external field, so the information about any early planetary field, which is stored on Earth and Mars, is lost on Venus.

produced a steady improvement in our knowledge of the surface features that lie hidden beneath the clouds of Venus. In the reasonably complete picture that we now have, an overwhelming feature is the presence everywhere of volcanoes. It has been estimated that there are more than a million volcanic features of various kinds on Venus. Some of them are huge mountains (Plate 17), others quite small domes, cracks and vents, and every scale in between is present. Many of them may be currently active, continuously emitting large quantities of water, carbon dioxide and sulphurous gases into the atmosphere.

The two major mountainous continents, surrounded by lowlands, are the most striking feature of a global map of Venus (Plate 9). Ishtar Terra is the largest, covering an area about the size of Australia, rising steeply from the plains in the far north of Venus.[4] The western part is a high plateau (3 kilometres above the mean radius of Venus) bordered by tall mountains that reach a further 3 kilometres in altitude. In the middle of Ishtar stand the Maxwell Montes, mountains that rise to 11 kilometres above the mean, higher and steeper than Everest, and the highest place on Venus. The pressure at the summit is 'only' 60 atmospheres, and the temperature more than 100 degrees less than the plains below.[5]

The nearest analogue to Ishtar on Earth is the Himalayas, in terms of appearance at least since they may have originated differently. The Himalayas were produced by a huge and energetic collision between surface plates trying to move sideways into each other, as part of the plate tectonic activity that still goes on all over our planet. The mountains, including Everest, will therefore continue to move and evolve, and are relatively short-lived in geological terms. Maxwell, on the other hand, more probably formed as a result of vigorous upwelling in a large convective 'hot spot' in the crust of Venus. The resulting massif is so very massive that it is debatable whether the plume must still be acting to support it, or whether the solidification and dehydration that followed uplift formed a sufficiently strong and rigid structure that it can support its own weight.

Stretching for about 10,000 kilometres along and south of the equator, Aphrodite Terra is the other very prominent highland region, in this case covering an area about equal to that of Africa, but with a more elongated shape that is reminiscent of a scorpion (Figure 4.7). The western end of Aphrodite is made up of two elevated, ancient, fractured plateaus. The highlands to the east of these extend for 5,000 kilometres and contain steep valleys, some of them on an enormous scale not found on Earth. The largest, Diana Chasma, is on a similar scale to the vast Valles Marineris on Mars; either would dwarf the Grand Canyon in Arizona. Smaller, but still prominent enough to have been one of the first continental-scale features discovered on Venus, is Beta Regio (Figure 9.3), which measures about 2,000 by 25,000 kilometres.

The lowlands on Venus, generally designated plains or planitiae, are the most feature-less regions in terms of having few tectonic and volcanic structures. The planitiae would have been the seabeds in ancient times when or if Venus was a water world like Earth. Whether or not that was the case, they certainly have been flooded by lava relatively

---

[4] Ishtar is at a latitude of about 70 degrees north, roughly where Iceland is on Earth. The most poleward part of Australia, near Hobart, Tasmania, is only 42 degrees south.

[5] On Everest, by comparison, the temperature falls to about -40 centigrade and the pressure to one-third of an atmosphere.

Figure 9.3 A *Magellan* radar image of a region 600 kilometres across near the south-eastern edge of Beta Regio, which contains the landing site of *Venera 10*. The position where the long-dead spacecraft sits cannot be determined exactly, but the surface panorama recorded when it landed looks more likely to be on the dark plain rather than the brighter irregular terrain (tesserae).

recently, in either a massive episode or some kind of quasi-continuous process. The question of how recently the vast floods of lava flowed is a hugely controversial one, probably requiring the return of samples to Earth for dating before it will be resolved.

On the surface: mountains, continents and river valleys

About 90 per cent of the surface of Venus is made up of features attributable to volcanic activity. The actual volcanoes that have been observed can be divided into three groups according to their size. The largest are the *shield volcanoes*, of which more than 100 have been identified in the *Magellan* maps, mostly in high regions lying 3 to 5 kilometres above the surrounding area. The mountains Maxwell (Figure 9.4) and Maat are the most prominent examples in the northern and southern hemispheres respectively.

Figure 9.4 A computer-rectified and 20× vertically exaggerated representation of the largest mountain on Venus, Maxwell Montes, based on *Pioneer Venus* and *Venera* radar data obtained from orbit.

Elsewhere, many of the smaller volcanoes are clustered together to form *shield fields* that can cover an area of more than 10,000 square kilometres. There are hundreds of these on Venus, some with extensive lava flows surrounding them, sometimes associated with regions where the crust has been cracked and broken up by tectonic activity.

The intermediate-sized volcanoes are subdivided and named for their appearance. The commonest are the *coronae*. These are large, circular patterns of ridges and troughs ranging in diameter from 75 to over 2,000 kilometres (Figure 9.5a). Although basically tectonic features, the coronae have their origins in volcanic activity below the surface, the crust being bulged and cracked by convection in the subsurface lava field, which may or may not break through the surface. There are various kinds of coronae, including some that may have been produced as a result of lava escaping and forming a mound, which then collapses, resulting in additional patterns of cracking and movement.

In some regions where the crust is particularly thin or weak, the cracks or *graben* that radiate out of some coronae often extend large distances, well outside the region of lava flow. About fifty of the type the *Magellan* investigators called *ticks* have been found. These are flat, circular volcanic domes about 25 kilometres in diameter with cracks or ridges radiating outwards, giving them an appearance reminiscent of the eponymous insect (Figure 9.5b). If the 'legs' are conspicuous but the 'body' is underdeveloped then they are termed *novae*, the assumption being that they are in the early stages of their evolution. Higher resolution images, and perhaps signs that changes are taking place over time, will

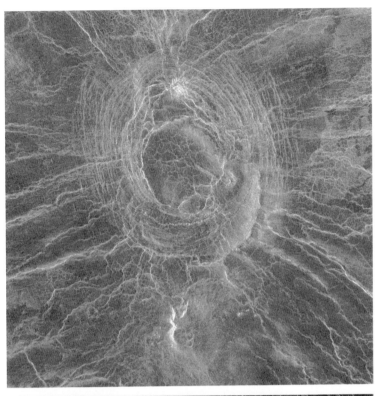

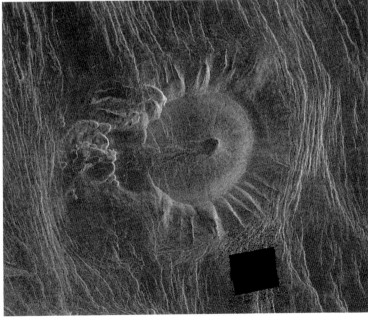

Figure 9.5 Top, a typical corona, about 200 kilometres across. The cracks are formed when a plume of lava pushes up on the solid crust from the liquid mantle below. Below, an example of the type of volcanic feature known as a 'tick'; this one near Alpha Regio is about 30 kilometres across. The dark spot near the centre is the caldera or crater from which the lava issued. The top is nearly flat, but with a raised rim; the 'legs' are produced by landslides in the steep cliffs which form the outer edge of the rim.

eventually reveal their true nature and where they belong in the 'zoo' of features that we really only glimpse in the *Magellan* pictures.

Examples of a different kind of modified corona are the *arachnoids*, volcanic mounds that have collapsed, cracking the crust and producing an insect-like shape (Figure 9.6a), and the *anemones*, which are relatively rare, with only 25 so far identified. The latter resemble the familiar sea creatures with 'hairy' flow patterns typically 50 kilometres across, radiating out from a central source of magma (Figure 9.6b). Again, a lot more detailed data will have to be gathered before we understand how and why these distinctive objects form.

Similar to ticks but without the 'legs' are the steep-sided, flat-topped volcanoes known as *pancake domes* (Figure 9.7a). These have well-defined circular outlines and patterns of radial fractures at the edge, and one or more small calderas near the centre. They often occur in groups, sometimes overlapping one another to produce complex shapes, such as the 50-kilometre-long 'mitten' shown in Figure 9.7b.

The pancake shape probably means that the lava that formed them had a higher viscosity than that found emanating from the more Earthlike large volcanoes, and so did not flow as freely or as far. The evidence from the *Venera* landers which touched down on the plains on Venus suggests that they are composed of basaltic material, low in silicates, similar to most terrestrial lavas. Were the pancake domes and related features made of something harder, more akin to the granite that makes up most of the ocean bedrock on Earth, this might have the properties required to explain the observed formations. We may have to land on them and sample the exuded material before we know for sure. The nearest analogues to pancake domes found on the Earth are at the bottom of the sea, where the density of the surrounding fluid plays a role in the cooling and solidification of the dome.

Sometimes lava is seen flowing from volcanic vents that are not associated with any kind of cone or dome. In this case the reason may be that they emit molten rock with an unusual composition that corresponds to a relatively low viscosity. If runny fluid is sometimes extruded this would also help to explain the many remarkable sinuous valleys on Venus, some of which extend many hundreds of kilometres from around the high volcanoes to the lava-filled flood plains (Figure 9.8). Many of them are deep enough to suggest that whatever flowed in them did so steadily for long periods of time, as in the canyons produced by rivers on Earth. In fact, they may have flowed more strongly as the results are longer, wider and deeper than the Amazon, the Nile or the Zambesi.

Obviously, running water was not involved in the case of Venus, but rather something that has a melting point that is less than but not too different from the mean surface temperature on Venus. Assuming this temperature has not changed greatly since the rivers flowed, the most likely candidate would seem to be a mineral such as carbonatite, which is sometimes expelled by terrestrial volcanoes.[6] Other materials cannot be ruled out, even including the low melting-point metals like lead or tin if they are available in sufficient quantity, although this is unlikely. Although the plains are clearly solidified now, and the

---

[6] Carbonatite is a mixture of minerals with a high carbon content and a low melting point (typically 500–600°C). It is found as a marble-like solid in some ancient lava flows on Earth, and there have been recent eruptions of at least one volcano (Ol Doinyo Lengai, 'Mountain of God', in northern Tanzania) where the lava flow was dominated by carbonatite.

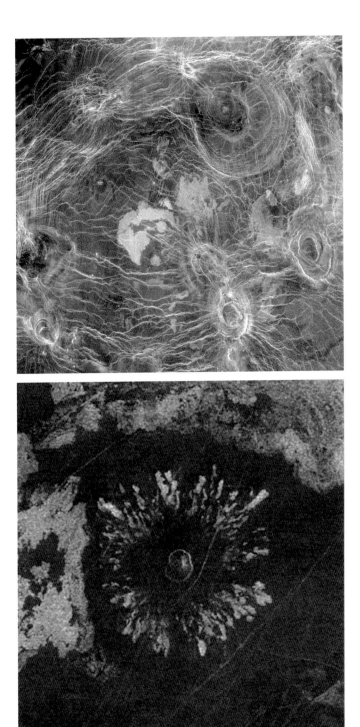

Figure 9.6 The weblike volcanic features in the upper image are arachnoids, 50 to about 200 kilometres in diameter. They are probably attempted volcanoes in which the lava pushed up and cracked the surface but did not erupt. Below is an example of an anemone, a small volcano which has thrown out lava in a flowerlike pattern. This one, about 40 kilometres across, sits on Atla regio,[7] a mountainous region just north of the equator.

---

[7] Atla is one of the Nine Mothers of the god Heimdallr in Norse mythology.

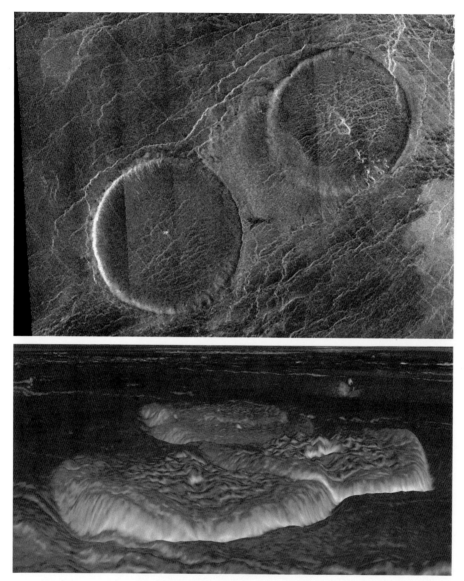

Figure 9.7 The 'pancake' domes, such as these about 60 kilometres in diameter (top) in Tinatin Planitia, are a different kind of volcano.[8] The difference appears to be due to the viscosity of the lava, that which forms pancakes being thicker and less mobile, so that flat tops and steep edges form. (Below) Looking like a big discarded mitten, this complex of pancake volcanoes is shown in a *Magellan* radar image that has been manipulated to give a three-dimensional perspective, including a twentyfold exaggeration of the vertical dimension. It is in fact only about a kilometre high but 50 kilometres across.

channels in places show evidence of being very old, we do not yet know for certain that rivers are not still running somewhere on Venus, especially if the fluid they contain is different in composition from the large-scale lava flows. Again, we need samples.

---

[8] Queen Tinatin founded a monastery in sixteenth-century Georgia near Telavi and is buried there.

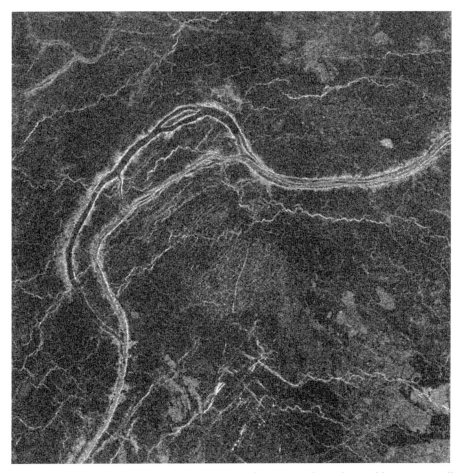

Figure 9.8 A segment, about 50 kilometres long, of a sinuous channel resembling a river valley, one of many on Venus. These have some of the characteristics of rivers on Earth; for instance, this one on the lowlands south of Ishtar, in the region called Sedna Planitia,[9] has formed a kind of oxbow lake.

Tectonic features

Tectonic features are those produced by movements of parts of the crust relative to each other. While Venus does not seem to have dynamic, continent-sized plates like the Earth does, there is still plenty of evidence for tectonic activity on a smaller but still sometimes quite grand scale. The existence of these is evidence that the dry, rigid crust on Venus cracks more easily than that on Earth, and it may be this property that prevents the formation of major plates by breaking them up more locally.

Among the results are the broad trenches bounded by cliffs, called *chasma*, where large sections of the crust have pulled apart (Figure 9.9) in response to movements in the

---

[9] In Inuit mythology Sedna is a goddess who lives at the bottom of the sea.

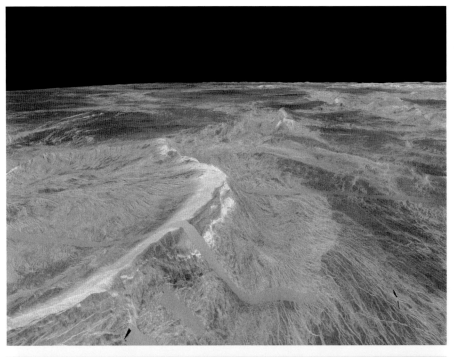

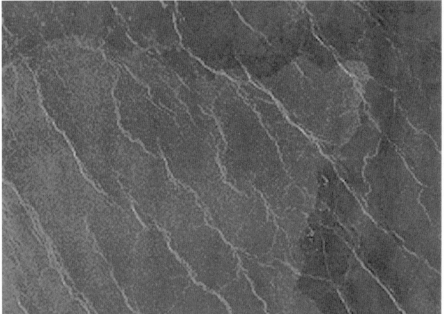

Figure 9.9 Large and small tectonic rifts seen in *Magellan* radar images. (Top) Dali Chasma[10] is a canyon 3 kilometres deep and thousands of kilometres long that forms the tail of the scorpion-shaped Aphrodite Terra (Figure 4.7). Latona Corona is on the left.[11] (Bottom) This part of Rusalka Planitia,[12] about 200 kilometres across, is typical of the wrinkled plains that are common all over Venus, formed when volcanic lava cooled and cracked after it filled the area.

---

[10] Venusian features are named for women, so Dali is not Salvador or the Lama but rather the goddess of the hunt, and seductress of the hunter, in Georgian mythology.

[11] Latona or Leto is the daughter of Phoebe and mother of Apollo and Diana in Greek legend.

[12] Rusalka is a Czech woodland sprite.

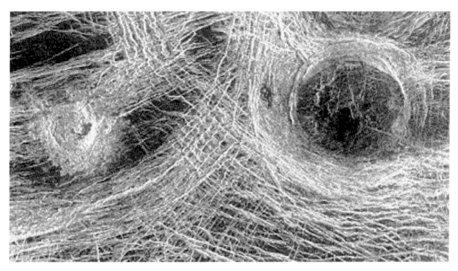

Figure 9.10 Complicated ridged terrain, produced by stretching and compression partly associated with the creation of the volcanic features, but possibly also other causes including surface temperature variations due to global climate change.

liquid mantle some 30 kilometres below. Finer-scale cracking is seen (Figure 9.10) in the flat volcanic plains that occur all over Venus, apparently produced as the lava cooled and solidified, quite recently in some cases. Elsewhere, the patterns are more complicated, with several different kinds of stress acting in a variety of directions, and perhaps in different epochs. Some of the complicated patterns in Figure 9.10, for instance, are clearly associated with lifting of the crust to form volcanic cones. The other patterns might even be the result of climate change, in which the temperature of the surface changed by enough over geological time periods to cause cracking due to thermal expansion and contraction.

Some of the networks of intersecting ridges and troughs (known as tesserae or 'tiled' terrain) were formed when local plates were forced together by movement of the crust, or by the lifting and subsidence of volcanic structures, including the large volcanoes themselves. Tesserae are overlaid by other features suggesting that they are among the oldest regions we can see on the Venusian surface. In addition to major earthquakes and volcanic upheavals, the high temperatures may be responsible for some of the disruption, especially if there have been large changes in the climate producing thermal stresses on a large scale all over the planet.

Impact craters

Relatively few impact craters can be seen on Venus compared with Mercury, Mars or the Moon, although all of them must have had a similar bombardment in the past. The older craters have probably been obliterated by atmospheric erosion, crumbling of the crust as a result of tectonics, and especially by the copious lava flows seen everywhere. There are almost no craters smaller than a kilometre or so in diameter, presumably because the thick atmosphere causes smaller or looser bodies to break and burn up before they hit the surface. Various dark splotches have been found on the surface, which are interpreted as the signature of meteoritic impacting material too fragmented to make a crater.

Impact craters are classified by their appearance into one of six categories. *Structureless* craters are the simplest, and most often the smallest, with a flat and featureless floor. *Central peak* craters have a central uplift that rises above the crater floor and terraced walls. They are particularly circular in outline and range in size from 8 to 80 kilometres, with most between 20 and 30 kilometres. *Double-ring* craters are typically greater than 40 kilometres, with an outer rim and an inner circle of peaks and ridges. The largest craters on Venus, ranging from 80 to 280 kilometres in diameter, are of the *Multiple-ring* type,

Figure 9.11 Impact craters are not very common on Venus because the thick atmosphere protects the surface from small or fragile meteorites, and also because most of the older craters have been obliterated by lava flows. The largest, named Mead,[13] is 280 kilometres in diameter and has a smooth floor (top), indicating that it filled with molten material after the impact. Below, a smaller (80 kilometre diameter) multi-ring impact crater, Mona Lisa.

---

[13] Named for the anthropologist Margaret Mead (1901–1978), who, according to Wikipedia, 'was a champion of broadened sexual morals within a context of traditional western religious life'.

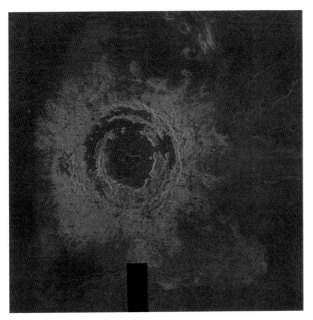

Figure 9.11 (cont.)

with two or more sets of concentric outer walls. *Irregular* craters, which have non-circular rims and broken-up but nearly flat crater floors, are very common. Almost a third of the craters on Venus are of this type, but most of them are quite small, 16 kilometres across or less. *Multiple* craters form when a falling body fragments into pieces, each creating a separate impact crater up to 45 kilometres in diameter and sometimes forming a distinct chain.

Less than 1,000 features have been found in the *Magellan* maps that can definitely be associated with impacts. This is not enough for the sort of statistical treatment that has been so useful for dating regions on the Moon and Mars, and in addition the size distribution is skewed by the tendency of the thick atmosphere to prevent smaller meteors from reaching the surface. However, most of the impact craters that are seen on Venus appear unmodified and therefore young (Figure 9.11), with a few exceptions (Figure 9.12). The rest have been covered over and in many cases must have been completely eliminated. This is the main basis for the theory that resurfacing on Venus takes place in isolated, vigorous episodes.

Atmospheric erosion by wind and rain has a major role in obliterating craters on Earth, but there does not seem to be any comparable process on Venus as most of the features in the *Magellan* images appear to be remarkably pristine. For one thing, the surface winds on Venus are much less than on Earth, and this is probably a more important factor than the high density. Perhaps surprisingly, the action of water and vegetation on the Earth seems to be a more effective force for corrosion of the topography than the sulphurous and other compounds which abound in the searingly hot air near the surface on Venus.

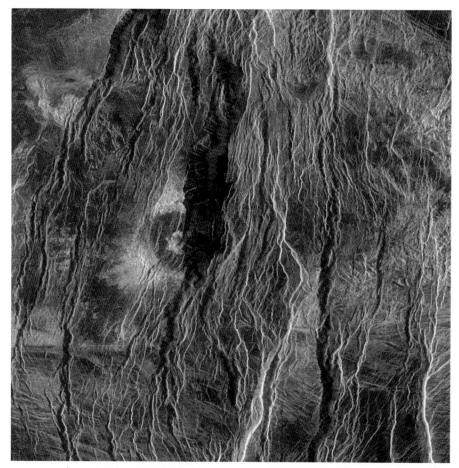

Figure 9.12 The 37 kilometre diameter impact crater named Balch[14] in the Beta region has faded and been partially obliterated by tectonic activity, suggesting it is ancient.

Lava plains

The infrared and radar maps have shown that most of the surface of Venus is covered with flat plains, like ocean beds except that they filled with lava which then solidified (Plate 18). Individual lava flows (Figure 9.13), some massive and some apparently very recent, are seen everywhere. The scarcity of impact craters is generally taken to be evidence that global resurfacing occurred on Venus about 500 million years ago. This catastrophic event is alleged by its advocates to have covered the low-lying part of the planet with around 10 kilometres depth of fresh lava, with massive outgassing of carbon dioxide, sulphurous gases and water vapour that could have had a large but temporary effect on the climate.

---

[14] Emily Balch was an American economist and Nobel laureate who died in 1961.

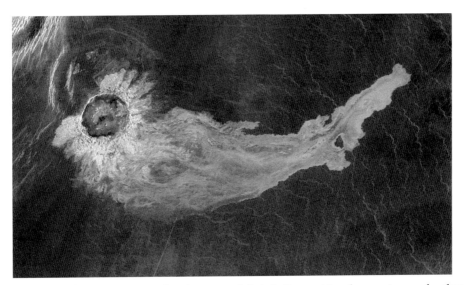

Figure 9.13 Some impacts punch a deep enough hole in the crust to release a stream of molten material, stimulating a volcanic eruption. In other cases, especially where the flow is more modest, it may be caused by the energy of the impact liquefying some of the crust. This *Magellan* image of Addams,[15] a 90 kilometre diameter crater in the Aino plain,[16] shows both splash deposits and a lava outflow stretching over 600 kilometres to the east.

Metallic snowcaps

Attempts have been made to identify the composition of the substance that has apparently fallen or condensed onto the mountains on Venus, as described in Chapter 4, with another example in Figure 9.14. The strongest clue comes from the temperature at the snowline where the material freezes. Of course, the temperature is much too high for water ice and the first suggestion was the metal tellurium. This has the right condensation temperature, but a closer look suggests that it is unlikely to occur as the pure element in sufficiently large amounts. A better bet might be the amalgam of tellurium and mercury with the chemical formula $HgTe$, which is more stable and has been found in various places on Earth.[17]

Another intriguing candidate is iron pyrite, known as fool's gold from its resemblance to the precious metal. Like the silvery metal tellurium, this fits the temperature and reflectivity/emissivity data, but the delightful vision of a planet of silver- or gold-capped mountains was dented by the planetary geochemist Bruce Fegley when he dismissed both possibilities on the grounds of implausible

---

[15] Jane Addams (1860–1935) was the first American woman to receive the Nobel Peace Prize, for her social work.

[16] Aino is a Finnish water spirit.

[17] Coloradoite, as the compound is known, was first discovered in the eponymous region of the USA just a year after Colorado achieved statehood in 1876.

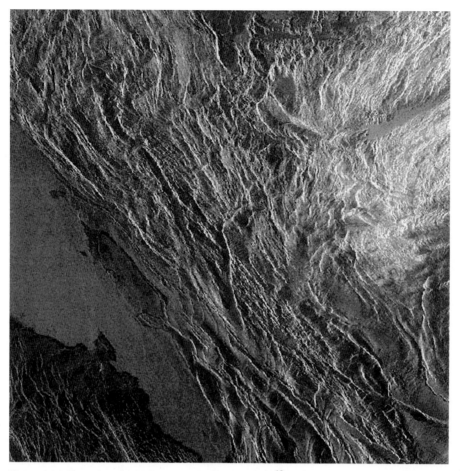

Figure 9.14 'Snow' on the highest part of Akna Montes,[18] on the western edge of Ishtar Terra.[19]

abundances, and arguments based on chemistry and dielectric constants estimated from radar data.[20] He and his colleagues proposed instead that the less exotic compound lead sulphide is the most likely candidate. Still, the view, and the compositional data, from a future probe that lands on one of these snow-capped peaks, is something to look forward to.

---

[18] Akna is the goddess of birth and fertility in Mayan and Inuit culture.

[19] Ishtar is the Babylonian and Assyrian goddess of love.

[20] Any tellurium, for example, should be present as tellurium sulphide, a gas, not plated out as metal.

Plate 1 An enduring figure in culture and mythology: Venus enthroned, by Henrietta Rae (1902).

Plate 2 Venus from the Earth, above as seen with the naked eye and below through a moderate-sized telescope (24 inch on Table Mountain, California).

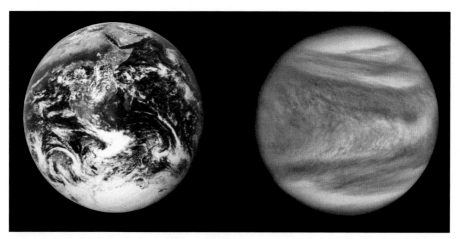

Plate 3 Globes of Earth and Venus to the same scale.

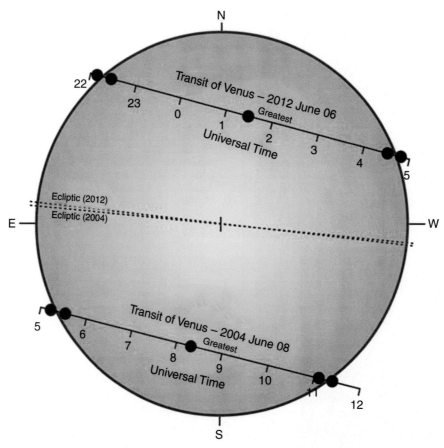

Plate 4 Transits of Venus took place in 2004 and 2012; the next one is not until 2117.

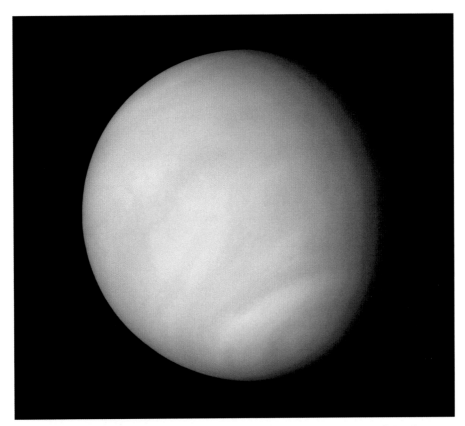

Plate 5 This is approximately what Venus would look like to the naked eye if the observer were approaching by spacecraft with the Sun to his or her back, so the disc is almost fully illuminated. It would be dazzlingly bright, however, and the faint dark markings would be even fainter and probably not discernible without a blue filter or other assistance.

Plate 6 Visions of Venus: top, after Arrhenius, *c.*1908, and the current view below, except that current thinking leads us to expect cloud-to-cloud lightning to be more common than cloud-to-ground, as depicted here.

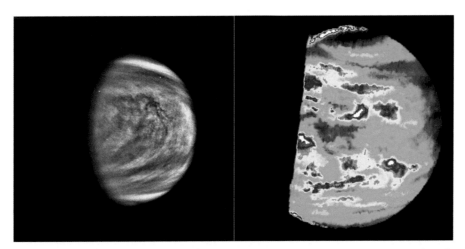

Plate 7 Two views of the disc of Venus from the *Galileo* spacecraft during its approach to the planet on 15 February 1990. The colours are artificial, used to emphasise that the left-hand image was taken through an ultraviolet filter, while the right-hand image of the nightside of Venus is in the 2.3 μm 'window', obtained by the near-infrared mapping spectrometer on *Galileo*.

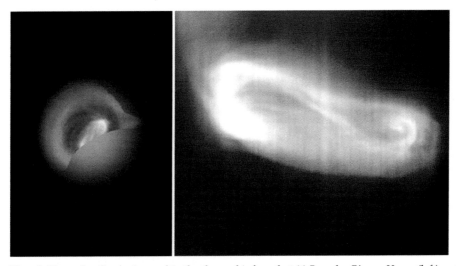

Plate 8 The polar dipole, imaged in the thermal infrared at 11.5 μm by *Pioneer Venus* (left) and at 5 μm by *Venus Express*. The scale of the bright 'eye' is such that it just fits inside a map of the continent of Europe on the same scale.

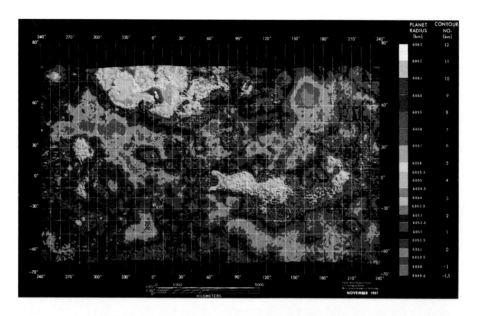

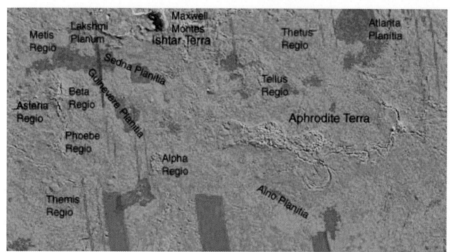

Plate 9 *Pioneer Venus* (top) and *Magellan* radar maps of the surface of Venus. The most prominent features are five extensive, high regions analogous to terrestrial continents, surrounded by extensive, low-lying plains. These are known as Ishtar (top, left of centre, with the very high Maxwell Montes at one end), the scorpion-shaped Aphrodite (lower middle, right of centre), Alpha (directly below Maxwell, just south of the equator), Beta (towards the left side of the map, above the equator) and Atla (to the right of Aphrodite, near the right edge of the image near the equator, and containing the high mountain Maat Mons).

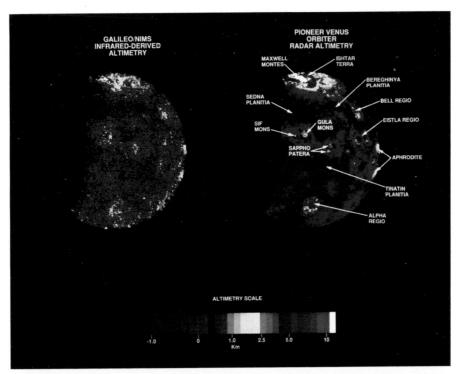

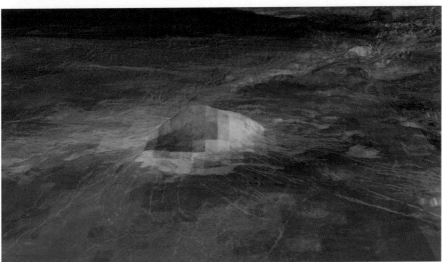

Plate 10 Surface features on Venus seen in the near infrared, at 1.05 microns wavelength, by the *Galileo* near-infrared spectrometer (top left), and by radar from *Pioneer Venus* (top right). Below, fresh lava on Idunn Mons observed by *Venus Express*.

Plate 11 Instead of using radar to bounce signals off the surface of Venus, the *Magellan* image at top uses the antenna to detect passive radio-thermal emission from a 2-kilometre-high volcano near Phoebe Regio. Note the difference between the reddish lava flows and the bluish high-altitude 'snowcaps'; red represents high and blue low emissivity. The image of a similar hilly region below uses *Magellan* radar data that have been processed to simulate the view as it might appear from a low-flying aircraft.

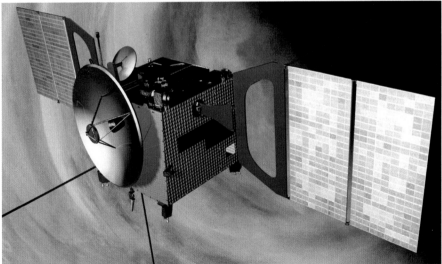

Plate 12 *Pioneer Venus Orbiter* in a clean room at Hughes Aircraft Corporation in Long Beach, California prior to delivery to NASA (top) and (below) an artist's impression of *Venus Express* in orbit.

Plate 13 The author (centre) with Astrium engineers at Stevenage, England, inspecting the *Venus Express* spacecraft under construction in 2004.

Plate 14 The Venus Express Science Team at a meeting in Kiruna, in the far north of Sweden. The ESA project's Chief Scientist, Håkan Svedhem, is centre right (hands clasped). The author is two places to his left. Note the Japanese (and American) participants.

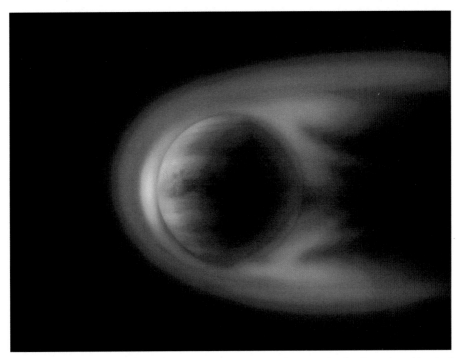

Plate 15 A visualisation of the Venusian magnetosphere showing the shock produced by the impact of the solar wind (blue) and water escaping as hydrogen and oxygen ions (yellow).

Plate 16 The *Venus Express* news conference at the ESA control centre in Darmstadt, Germany, at the time of its arrival at Venus. L to R: the author, Dr Dimitri Titov, the TV presenter, behind a one-tenth scale model of the spacecraft.

Plate 17 Maat Mons, the second-highest peak on Venus, in a *Magellan* radar image from orbit reprocessed to give a surface perspective. The vertical scale has been vastly exaggerated; the volcano is eight kilometres tall and about four hundred kilometres in diameter at the base.

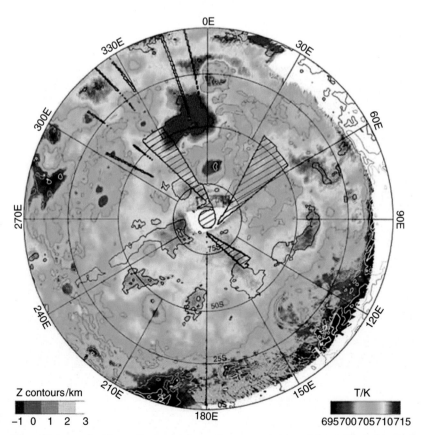

Plate 18 A map by *Venus Express* of the surface temperature of the southern hemisphere of Venus, superimposed on height contours from *Magellan*. Most, but not all, of the apparent temperature anomalies are due to topography.

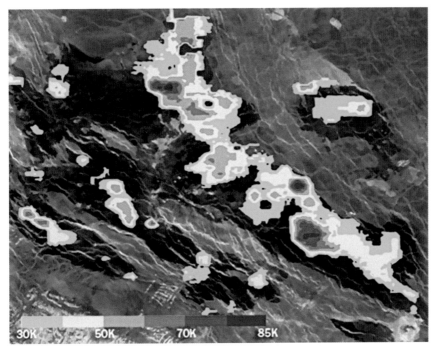

Plate 19 Temperature anomalies in Bereghinia Planitia seen in close-up in *Magellan* microwave emission measurements, superimposed on a radar image of the same area.

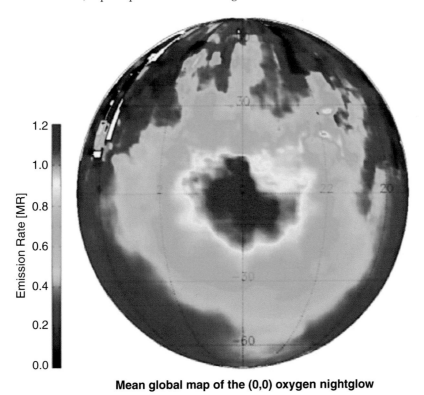

Mean global map of the (0,0) oxygen nightglow

Plate 20 A map of the nightside showing the brightness of the oxygen airglow at 1.27 microns wavelength, measured by *Venus Express*. Atomic oxygen is produced by solar radiation on the dayside of the planet, and is transported to the nightside where it recombines and produces the glow, with a strong maximum near local midnight at the centre of the image.

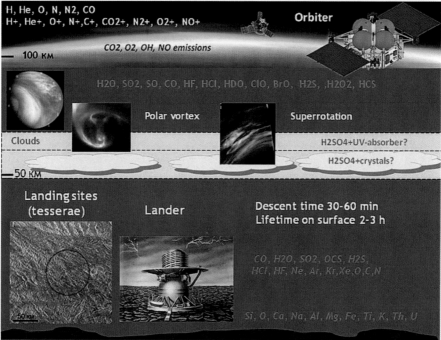

Plate 21 *Venera-D* is scheduled to be the next Russian mission to Venus, but has been repeatedly delayed. Top, an artist's impression of the spacecraft approaching Venus. Below, a montage illustrating the scientific objectives.

Plate 22 A prototype of a balloon designed to float on Venus, seen here in the laboratory at the NASA Jet Propulsion Laboratory.

Plate 23 The *Venus In-Situ Explorer* floats in the deep atmosphere, just above the surface.

Plate 24 A solar-powered aircraft may one day soar above the clouds on Venus. More advanced versions could fly below the clouds, for high-resolution imaging of the surface.

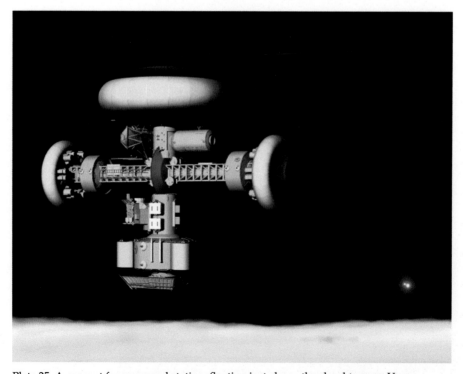

Plate 25 A concept for a manned station, floating just above the cloud tops on Venus.

Chapter 10
Atmosphere and ocean

The pre-space age planetary astronomers knew that the albedo (reflectivity) of Venus is higher than that of Earth, which must partially offset the extra heating that comes from being closer to the Sun. It was widely expected that Venus would turn out to be a more tropical version of the Earth, but no one pictured a climate as extreme as the reality that was first detected by ground-based radio astronomers and confirmed by *Mariner* 2 and *Venera* 9. This was a big surprise at the time.

In fact, the modern value for the albedo, integrated over wavelength, is more than two and a half times that of Earth, at about 76 per cent rather than 30 per cent, so that Venus actually absorbs less radiative energy than Earth, despite the Sun appearing twice as large in the sky[1]. Thus, it could be argued that Venus should be not warmer but actually cooler overall. This, however, does not take into account the huge difference in atmospheric thickness. On Venus the surface temperature enhancement through the blanketing 'greenhouse' effect is similar to, but more than ten times larger, than the corresponding effect on Earth.

If we compare the temperature profiles on the two planets as a function of atmospheric pressure, rather than height above the surface in the more usual way, they look a lot more similar (Figure 10.1). The mean temperature on Venus is, in fact, only slightly higher than that on Earth if the comparison is made at the atmospheric pressure that is characteristic of the surface of the Earth. The biggest difference is the bulge in Earth's stratosphere produced by heating in the ozone layer, a phenomenon not found on Venus.

What the early space probes discovered was that Venus's atmosphere continues on down below the 1 bar level to higher and higher pressures, which forces the temperature up as well in a way that is predicted by simple physical laws. The solid surface is not reached until the pressure is nearly 100 times that on Earth, and the temperature there is nearly 450 centigrade. The massive amount of carbon dioxide which is responsible was exhaled from the interior and is still being topped up by volcanoes. Something similar happened on Earth, but this planet hung on to enough liquid water to turn the carbon dioxide into calcium carbonate and other solid minerals that bound the gas into the surface and cooled the climate to the habitable state we now enjoy.

---

[1] By area. The ratio of the distance to the Sun for the two planets is close to the square root of 2, so to an observer on Venus the Sun is about 40 per cent larger in diameter. (It would actually be possible for someone on the surface of Venus to see the disc of the Sun, dimly, at times when the clouds overhead are relatively thin.)

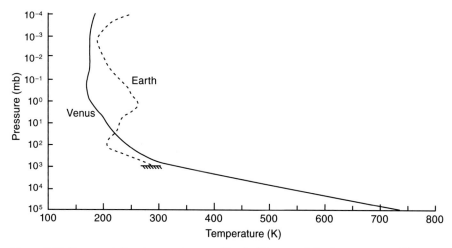

Figure 10.1 Representative temperature profiles for Venus and Earth, as measured by instruments on *Pioneer Venus* and *Nimbus 7* respectively, on a common logarithmic pressure scale. The peak in the stratosphere on Earth is due to heating by the absorption of sunlight in the ozone layer at ultraviolet wavelengths; without that, the profiles would be even more similar.

This broad picture is largely accepted by the scientific community, but is based on fairly limited evidence and there are a lot of uncertainties and loose ends. The timescale over which the climate changed, for one thing. The stability of the current state, for another: is the climate of Venus still changing? If so, where will it end up? Venus may have been an ocean world in the past, and may cool in the future and wind up being an attractive venue for astronauts and migrants from Earth. The reasons for the current dramatic differences between two planetary environments that, for good reason, were originally expected to resemble each other quite closely, have been partially provided in the past few decades, but there is much still to learn and understand.

Not-so-identical twins

Figure 10.2 shows a simplified, average profile of temperature versus height for Venus. The names of the regions are taken from those for Earth, since similar processes shape them both. The troposphere is the convectively overturning region nearest to the surface, where the opacity of the atmosphere is so high that radiative transfer of energy has a limited role. Above that, the stratosphere is the region where the pressure is low enough that radiation takes over the dominant role in heating and cooling. Its division into upper and lower stratosphere is fairly arbitrary on Venus because there is no local temperature maximum due to ozone, unlike Earth. However, it allows us to distinguish between the cloud-top region where photochemistry and the aerosol opacity it produces are important, and the nearly isothermal true stratospheric layer above that. In the thermosphere, the temperature rises with height because of the absorption of short-wave ultraviolet radiation from the Sun, and the low density and consequent small heat capacity. The exosphere

142

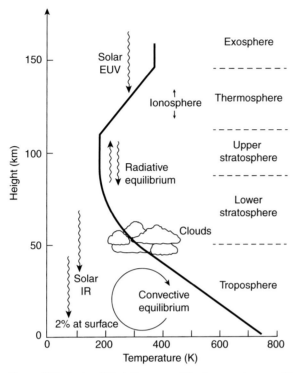

Figure 10.2 A simplified diagram of the temperature profile in Venus's atmosphere, also showing the major processes at work and the approximate locations of the main cloud layers.

is the very low density region where collisions between molecules are rare and escape to space is possible.

The composition of the atmosphere as we now understand it was presented in Chapter 6, where most of the key measurements on which the data are based are discussed. We saw that it consists of gases that are familiar to us from our own atmosphere, although in different proportions as well as different total amounts. The high pressure, and the ability of gases such as carbon dioxide and water vapour to absorb heat radiation, combine to warm the surface through an extreme version of the same greenhouse effect that maintains, and threatens to change, life on Earth. This was confirmed when it was shown that the same basic theoretical models can predict the temperature profiles that we observe on both Earth and Venus. This is comforting, to say the least, when we depend so much on these same models to predict the future of climate change on Earth.

It is worth saying again the fact that the main reason Venus is so much hotter at the surface is just that there is so much more atmosphere. The pressure, and hence the temperature, both continue to rise with depth below the 1 bar level because the profile must follow the hydrostatic and adiabatic formulae that follow from basic undergraduate-level mechanics and thermodynamics.[2] These formulae predict to a

---

[2] See the author's book, *Elementary Climate Physics* (Oxford University Press, 2005), for details.

temperature increase of about 10 degrees for each kilometre of depth below the 1 bar level. This amounts to about 450 degrees altogether at Venus's surface pressure of 92 bars. The temperature at the 1 bar level on Venus is not much different from the mean value on Earth at that pressure (i.e. Earth's surface), actually about 50 centigrade, making the predicted temperature at the surface of Venus the sum of these two, or 500 centigrade, close to what the *Veneras* measured when they landed.

Figure 10.3 shows examples of the match between some typical measured temperature profiles for both Venus and Earth and the predictions of simple radiative–convective model calculations. The agreement is not exact – it would be surprising if it were, since these models only consider the most basic physics – but close enough to confirm the assertion that the processes at work are basically the same in both cases. It follows that, unlike many other aspects of the climate on Venus, the basic hotness of the surface should no longer be considered a major mystery (how it got that way is another question). For now, we stress that the factor that was so surprising when it was first discovered is just a

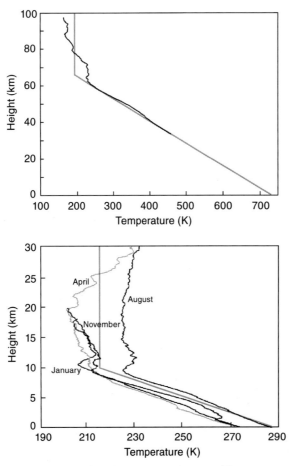

Figure 10.3 Simple radiative–convective equilibrium temperature profile models for Venus and Earth (grey) compared with representative measured profiles (black).

consequence of the large mass of the atmosphere, rather than any mysterious thermal process.[3]

The difference between Earth and Venus in terms of surface pressure may not be too surprising either, provided we can account for the history of water on Venus. Before we attempt to do that, let us take a closer look at the details of the Venus greenhouse effect.

Atmospheric composition and the greenhouse effect

Entry probes like *Pioneer Venus* measured the intensity of sunlight at various heights as they descended through the atmosphere and down to the surface. They found that enough sunlight diffuses through the cloud layers on Venus to provide a moderate amount of heating at the surface. This is only about 12 per cent of the total solar heating of Venus, the rest of the energy being absorbed at various levels in the atmosphere. However, that translates to just under 3 per cent when we allow for the fact that around three-quarters of the incoming sunlight is reflected back to space (see Figure 10.4, where these estimates are compared to the equivalent numbers for the Earth).

The energy that is deposited at and near the surface is trapped, in the sense that it cannot escape as radiation back to space, because of the opacity of the overlying atmosphere. This is much higher for heat radiation than it is for sunlight because the absorption

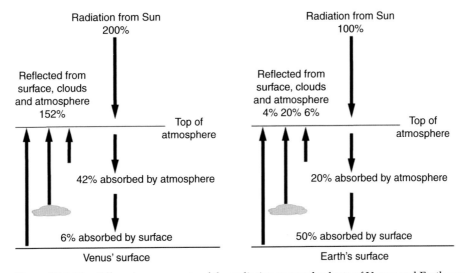

Figure 10.4 The different components of the radiative energy budgets of Venus and Earth are shown as planetwide averages, taking the solar irradiance at the Earth as 100 per cent and Venus as twice that.[4]

---

[3] There is still a following on various websites for the renegade Russian scientist Immanuel Velikovsky, who maintained in *Worlds in Collision* (1950) that Venus is hot because it is a young world, still cooling after its ejection from Jupiter (*sic*), an event that he found recorded in the Old Testament.

[4] Actually, the sunfall at Venus relative to that at Earth varies between 182 per cent and 200 per cent of Earth's when the orbital eccentricities of the two planets (0.007 and 0.0167, respectively) are taken into account.

bands of CO_2 and other gases and cloud particles are much stronger at the longer wavelengths. Instead, the surface is cooled by convection involving rising parcels of warm air.

Convection is not very efficient compared with radiation, so the surface gets hotter and hotter until the incoming solar flux is balanced by the heat brought by convection to a level near the cloud tops, where it can at last radiate to space. An airless body with the same albedo and distance from the Sun as Venus would be both heated and cooled by radiation, and, despite the fact that a much larger fraction of the Sun's radiation would heat the surface, calculations tell us that it would be in equilibrium with radiative cooling for a mean surface temperature as low as −45 degrees centigrade. Thus, the atmospheric greenhouse adds around 500 degrees to the surface temperature, compared with just 35 degrees on Earth.

This calculated temperature is close to the actual value at the Venusian cloud tops, indicating that this is the level where most of the energy is radiated away. Global measurements by the *Pioneer Venus* orbiter of the net infrared emission and the total reflected solar energy confirmed that the planet is in overall energy balance to within the accuracy of the measurement. Although there was by this time little doubt, this further vindicated Carl Sagan's theory for why Venus is so much hotter than the Earth: it is just a more efficient greenhouse.

There is also the fact that more than 96 per cent of the lower atmosphere of Venus is composed of carbon dioxide, rather than nitrogen as on Earth. It is sometimes said that this is the main reason why Venus is so hot, since CO_2 is an important greenhouse gas (i.e. it absorbs strongly in the infrared spectrum), whereas nitrogen is not. This is true, but in fact it is a secondary consideration. As we have seen, the pressure is what dominates the difference between the two planets, and the extra opacity generated by carbon dioxide makes an important but relatively small difference. If the Earth had such a high surface pressure, it too would be extremely hot, even without the increased proportion of carbon dioxide that is found on Venus.

Energy balance and entropy

The average energy balance of the whole planet is, of course, only part of the story. Consider Figure 10.5, which shows how the total energy absorbed and emitted as radiation by Venus and Earth depends on latitude.

As we might expect, both planets are heated more than they cool near the equator, while the opposite is true near the poles. For the whole planet to conserve energy, the atmospheric (and, on Earth, oceanic) 'heat engine' must move energy from low to high latitudes. A lot of energy is involved – several petawatts.[5] This is the climate at work.

On Earth, about half of the work is done by the oceans, which move slowly but have a much larger heat capacity than the atmosphere. On Venus, the dense atmosphere must do it all. Data such as those in Figure 10.5 are telling us how the atmosphere circulates in order to keep Venus onside with regard to the most important principle in all of physics, the first law of thermodynamics dealing with conservation of energy.

---

[5] 1 petawatt is a million billion (10^{15} or 1,000,000,000,000,000) watts.

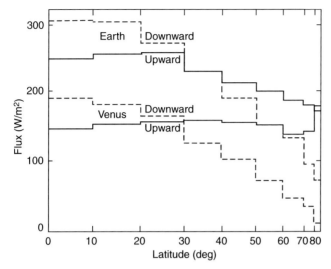

Figure 10.5 The energy balance of Venus versus latitude, as measured by the infrared radiometer on the *Pioneer Venus* orbiter, compared with similar data for the Earth.

Not far behind is the second law, which has to do with the more difficult concept of entropy.[6] General circulation models can be programmed on computers to simulate the motions in Venus's atmosphere, and show (or rather, be adjusted until they do show) that the transport of heat gives agreement with the measurements in Figure 10.5. The second law says that the work done must balance the entropy budget as well, but in preliminary models, which are all we have at the moment, it doesn't. When we look at the numbers, a component seems to be missing in the entropy budget, meaning that the contribution of a key process has been underestimated, or omitted. This probably has something to do with absorption and emission of radiation, and related dynamical activity, in the deep atmosphere below the clouds. This region remains mysterious, and even techniques, let alone real missions, for its systematic exploration remain a distant prospect.

Why so much CO_2?

We saw in Chapter 6 that the composition of Venus's atmosphere has been measured to find mostly carbon dioxide with a few percent of nitrogen, a small proportion of noble gases, principally argon, and small but significant amounts of more variable species including water and sulphur dioxide (Figures 6.6 and 6.7). At 3.5 per cent, the abundance of nitrogen is interesting – it looks small, but when we remember that there is 90 times as much mass of atmosphere on Venus, it follows that there are about three times as many

---

[6] Entropy was introduced as a mathematical and philosophical concept about 150 years ago by Clausius, and has seen great service ever since in studies of theoretical thermodynamics by luminaries such as Boltzmann and Gibbs. The usual simplistic definition is that it is a measure of the disorder in a system, which must increase with time if the system is closed, in accord with the second law. This can be considered a (or even *the*) definition of time itself.

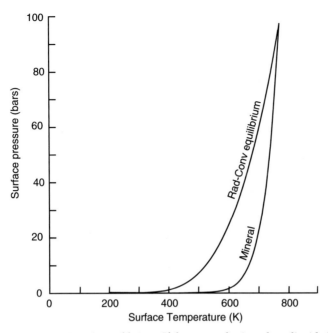

Figure 10.6 Urey's equilibrium. If the atmospheric carbon dioxide is in thermochemical equilibrium with the surface, and the surface is composed mainly of the common minerals calcium carbonate, calcium silicate and silica, then the surface pressure must lie along the line labelled 'Mineral'. Radiative–convective equilibrium is expected in the atmosphere, so it follows the other curve; both criteria are satisfied where the curves cross. The high temperature crossing is at the surface temperature and pressure found on Venus, while the low temperature one is close to the values on Earth.

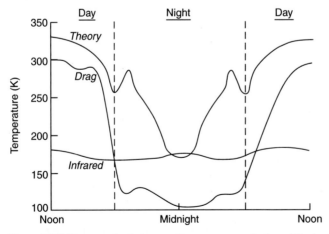

Figure 10.7 Thermospheric temperatures, measured at an altitude of 115 kilometres by infrared sounding and 40 kilometres higher by satellite drag measurements, both by *Pioneer Venus*. The third curve is a theoretical model of the expected temperature at an altitude of 150 kilometres.

nitrogen molecules in Venus's atmosphere as there are in Earth's. If we could take most of the CO_2 away from Venus, we would have an atmosphere somewhat more massive but not as hugely different from Earth as it is now. What would the surface temperature be then?

Before answering that, we should look in more detail at *why* there is so much more carbon dioxide on Venus relative to Earth. In fact, a plausible explanation for the apparent superabundance is not particularly difficult to find. The carbonate rocks on the Earth (coral atolls, the chalk in the white cliffs of Dover and all the other such places) formed from CO_2 that was once in the atmosphere. Geologists have estimated that they hold about the same amount of carbon dioxide in solid form as Venus does in its atmosphere. The conversion of gaseous CO_2 to solid carbonate occurs much more efficiently in the presence of liquid water, in which the carbon dioxide first dissolves, so it is the relatively water-depleted state of Venus which is most likely to be responsible for so much of the gas remaining in the atmosphere.

With any liquid water that Venus may have once had now long gone, it is tempting to think that the current surface pressure there is stable. However, it is well known that the CO_2 abundance in Earth's atmosphere can vary because of natural and anthropogenic factors, and that it is changing at the present time, with likely consequences for the global climate. If there is an equivalent force for climate change on Venus, it is probably also the carbon dioxide amount, although the cause is more likely to be long-term variability in the level of volcanic activity, rather than the things that concern us here such as man-made pollution and decimation of forests.

If, on the other hand, the climate on Venus is in a stable state in the long term, then it is likely that some mechanism stabilises the atmospheric carbon dioxide content. Harold Urey famously proposed in 1952 that the exchange between atmospheric CO_2 and common minerals likely to be plentiful in the surface of Venus may provide such a buffer.[7] The reaction he thought would dominate was the combination of calcium silicate with carbon dioxide to produce calcium carbonate (chalk) and silica (sand).[8] The reaction is reversible and has been shown since Urey's time, with the relevant laws of hydrostatics and thermodynamics, to reach equilibrium at precisely the temperature and pressure found on the surface of Venus (Figure 10.6). It is exquisitely tempting to assume that this remarkable fact solves the problem. Alas, few things are quite that simple in planetary science.

One of the several problems that have been raised with this theory is the question of how a sufficiently intimate contact between atmosphere and the minerals in the lithosphere is achieved. It is easy to show that the amounts exposed on the surface would not be enough to make the process work. Something to do with the tectonic activity that breaks up the surface may be the answer, perhaps even global plate tectonics if this is present after all. There is also the active volcanism and the relatively recent global resurfacing, which speak of an intimate relationship between atmosphere and interior.

---

[7] Harold Urey was an American physical chemist, most famous for his work on isotopes. His Nobel Prize in 1934 was for isolating deuterium and heavy water, making possible the studies of D/H in planetary atmospheres, including Venus.

[8] That is, $CaCO3 + Si02 \leftrightarrow CaSiO3 + CO2$

Even worse however, the equilibrium state that falls so attractively at the value we want turns out under further study to be unstable against any perturbation in the temperature or pressure at the surface, like a coin standing on edge. If Venus's climate has been the same for a long time, some additional process is at work to keep it there. Again, this is not inconceivable, although we cannot say that we understand it very well at present. The calculation of equilibria in complex systems is difficult even in a laboratory set-up, and not likely to be elucidated for the climate on Venus for a long time to come. What we do know at present is that a chamber containing calcium carbonate, silica and calcium silicate in an atmosphere of carbon dioxide at a pressure of 92 bars, is at equilibrium at a temperature of 450 centigrade, the same temperature a radiative–convective equilibrium model of the atmosphere finds, and the same as we measure, on present-day Venus (Figure 10.6).

Venus's primordial ocean

Apparently Venus kept most of its original inventory of CO_2 in the atmosphere because it had no surface water (or organisms) to dissolve it and turn it into coral and chalk. But why is Earth so wet and Venus so dry? Did they start out that way, or did they evolve along different paths, so that Venus used to have as much water as Earth, but lost most of it? The answer to this question is central to any understanding of Venus's remarkable climate.

Probably, Venus had water and lost it. It is hard to reconcile modern theories about the formation of the terrestrial planets with one planet that formed dry while its neighbours Earth and Mars acquired large amounts of water during their formation. It can be argued, of course, that as it is closer to the Sun, water might have been less easily retained by the hotter proto-Venus as it accumulated. However, it is considered more likely when all of the other evidence is taken into account that the formation of large planetesimals first in quite chaotic orbits would have mixed up any large variations due to distance from the Sun before the final formation of the planets we have today.

This still does not necessarily imply that all of the inner planets were wet. Earth and Mars could have formed dry and then at some later stage received a huge mass of water from one (or more, but not too many, or the statistics start to favour wet Venus again) large ball of ice (a comet) falling from space. Small comets, up to the size of a car or a house, rain down onto all of the planets all of the time, even today, and were probably more numerous in the past. The flux of these ought to be about the same, averaged over large numbers and long times, for all of the inner planets. But it is not too hard to imagine that Earth happened to receive one of the very rare big comets, thus endowing it with extra water the others did not get.

If this was a single event it would have to have been a very big comet – equivalent to more than a million of the largest ones we see today (5 million Halleys, for example), but this is not impossible. Some of the largest icy bodies in the Kuiper belt beyond Neptune could deliver this much water in one event if they strayed into the inner solar system, as they may well have done in earlier times. It is estimated that the collision on Mars that produced the giant basin we now call Hellas could have delivered enough water to cover the planet to an average depth of the order of 10 metres. Maybe Venus was just unlucky.

Furthermore, we expect that Earth, Mars and Venus accreted from material that contained water, not just as ice but also in rocks as water of crystallisation. There was a lot of this – enough to make Earth's oceans many times over, and we still see some of it coming from the interior as steam in volcanic vents of various kinds. We don't know at present what proportion of the water in the oceans came from outgassing, but if the answer is most of it, the amount should have been much the same on Venus. Then we need to consider where it went.

The atmosphere of Venus, including the clouds, currently has about 100,000 times less water than the oceans and atmosphere of the Earth. On both planets (and on Mars, in comets and elsewhere) this is a mixture of ordinary (H_2O) and heavy (HDO) water. The D stands for deuterium, the form of hydrogen that has essentially the same chemical properties but twice the mass of H, due to the presence of an extra neutron in the nucleus.[9] It has long been realised that measurements of the D/H ratio have great significance in theories of the history of water on the terrestrial planets. As we have seen, remote sensing and *in situ* measurements gathered from many different experiments point to vastly different D/H ratios on Venus and Earth, with Venus having more than 100 times the terrestrial value.

This relative overabundance of deuterium on Venus is considered by most planetary scientists to be the smoking gun that proves that Venus lost large amounts of water over the first billion years or so of its existence. The loss processes involve dissociation to form ions of hydrogen, deuterium and oxygen, which can then escape into space. The lighter hydrogen atoms escape faster, leaving deuterium enriched.

The rate of escape of hydrogen depends strongly on the abundance of water in the middle atmosphere. On Earth, the low temperature at the tropopause, about 10 kilometres above the surface, provides a cold trap that effectively creates a barrier for water trying to propagate up from the lower atmosphere and surface. Conditions on early Venus evidently were efficient at forcing water into the stratosphere, where solar ultraviolet photons readily dissociate the H_2O and HDO molecules thus creating favourable conditions for hydrogen escape. Theoretical calculations show that, with certain reasonable assumptions, Venus could have lost an ocean of present-day terrestrial proportions in this way in less than 500 million years. Others, equally reasonably, have estimated that an ocean may have persisted for more than 2 billion years, half of the lifetime of the planet.

D/H on Venus might be even larger today if all of the deuterium originally in the primordial Venusian ocean had been retained. Some of the deuterium must have escaped from the atmosphere, as well as hydrogen, but what proportions are to be expected? And what happened to the oxygen?

Attempts to model the early evolution of Venus in order to fill in the details of the loss of water run in to a number of difficulties. For instance, we do not know what the atmospheric pressure was then. If liquid water was abundant, the pressure could have been lower than now because CO_2 from the atmosphere would dissolve in the water and promote the formation of carbonate minerals, as happened on Earth. Also, it is likely

---

[9] Strictly, we have to include D_2O, HTO, T_2O, DTO and other isotopes in the definition of heavy water, but these are extremely rare and have not yet been observed on Venus. T is tritium, the hydrogen isotope with mass number 3. Oxygen also has rare isotopes that are heavier than the commonest one with mass number 16 and can occur in water, rendering it 'heavy'.

that plate tectonics would have been more active on an early wet Venus, with subduction and recycling of the crust also providing a sink for atmospheric gases. The lower atmospheric pressure would mean faster upward transport of H and D atoms, providing a pathway for escape. After the elimination of most of the water, carbon dioxide could build up in the atmosphere again, with volcanic activity continuing and no liquid water to remove it.

Water on Venus now

Today, the only water we find on Venus is in the atmosphere, as vapour or bound into cloud particles. The measurements we have, particularly those acquired by entry probes at altitudes below the clouds, have produced very conflicting results. Whether this is from measurement error, natural variability or both is not always clear.

The *Pioneer* and *Venera* gas chromatographs indicated concentrations larger than 1,000 parts per million near 40 kilometres, while their mass spectrometers first reported a constant water concentration around 100 parts per million between 10 and 25 kilometres, decreasing to just 20 parts per million at the surface. The scanning spectrophotometers on *Venera* also indicated a strong decrease in the water abundance with height, from 150 parts per million at 40 kilometres to 20 parts per million below 10 kilometres. These results potentially have important implications for the presence of sources or sinks of water at the surface, but are subject to many uncertainties and interferences, and they have been regularly revised and updated.

Unfortunately there have been no new *in situ* measurements since the *Vega* probes in the mid-1980s. It was suggested at the time that the direct sampling experiments were unreliable because of contamination by water in the probe itself. The spectroscopic data from the same probes, which might not have this problem, could be interpreted with a constant H_2O mixing ratio of about 30 parts per million below the clouds.

This much smaller value would be more consistent with modern spectroscopic data obtained through telescopes on Earth. For instance, spectra of the nightside collected with the Infrared Imaging Spectrometer on the Anglo-Australian telescope in July 1991 showed a water vapour mixing ratio of 20 parts per million at the base of the lower cloud, increasing to 45 parts per million at the 10 atmosphere level, then remaining constant down to the surface. The spectrometer orbiting on *Venus Express* has found near-constant values of around 30 parts per million at all heights below the clouds, and this seems to be fairly constant across the globe.

At greater heights, within and above the clouds, there still seem to be huge variations, linked to cloud formation and dissipation processes and some apparently vigorous meteorological activity in Venus's atmosphere. Most of this remains poorly understood, as we discuss in Chapter 13.

Exosphere and escape

100 kilometres or so above the surface, the nature of the atmosphere changes dramatically. The circulation switches from predominantly equator-to-pole, with the rapid zonal winds

superposed, to a subsolar to anti-solar regime flowing around the equator and over the poles. The most energetic solar radiation in the ultraviolet is absorbed at these levels and so does not penetrate much lower. It dissociates and ionises the atmospheric gases, producing high temperatures from the energy released.

The physics of this low-pressure region, containing only a small fraction of the mass of the atmosphere, would not be of great interest except to specialists were it not for the fact that this is also where gases can escape to space. Investigating, at least superficially, what is going on in the thermosphere and exosphere is crucial, therefore, for understanding how Venus maintains its high surface pressure and how it lost most of its water, and perhaps also some of the nitrogen and the carbonic and sulphurous gases as well.

The highest-altitude measurements of temperatures by infrared sounding are at about 115 kilometres above the surface, so they give relatively little information about the thermosphere. The altitudes probed by drag measurements on orbiting spacecraft, giving density information that can be converted to temperatures, extend down to an altitude of about 150 kilometres (or a little lower with some risk to the spacecraft, since too much drag can pull it out of orbit).

Temperatures from both of these techniques are shown in Figure 10.7 as a function of local time of day, and dramatic differences are apparent. Whereas there is hardly any day–night variation at the top of the mesosphere, with an almost constant temperature near −100 degrees centigrade, in the thermosphere the day time is much hotter, around 30 centigrade, and the night nearly 200 degrees colder. The night-time thermosphere on Venus is sometimes called the cryosphere for this reason.

Various other puzzles are evoked by these data. First, the higher altitude temperatures have a minimum at local midnight, whereas 35 kilometres lower down there is a weaker maximum at this time. Secondly, although this is not shown in Figure 10.7, which gives the long-term average temperatures over many days, the night-time temperatures at both levels oscillate by a few degrees with a period of about four days. This is probably the result of disturbances caused by the cloud-top circulation, which has a similar period, flowing over the obstacle produced by the subsolar convective disturbance and producing waves that propagate upwards. We would expect the phase of the wave to change with height, and it does: maxima at 150 kilometres corresponding to minima at 115 kilometres, and vice versa.

However, as is clear from Figure 10.7, the theoretically predicted temperature at the altitude of the drag measurement is very different from that measured. It falls much more slowly during the transition from day to night and back again, and reaches only about −100 degrees centigrade at the midnight minimum instead of the observed −160 degrees. Some understanding of what is going on can be gleaned from Figure 10.8, which shows the results of calculations of the heating and cooling rates in this part of the atmosphere. The cooling, in particular, is strongly peaked around 125 kilometres, a property of the carbon dioxide that still makes up most of the atmosphere at these levels. But some quite complicated dynamics are required as well, to explain the very sharp fall of nearly 200 degrees in the temperature across the day–night boundary.

The theoretical model, as we might expect, predicts strong winds driven by pressure differences across the day–night boundaries in the thermosphere, which tend to reduce the

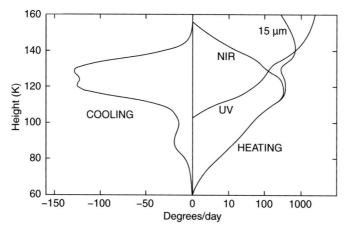

Figure 10.8 Calculated heating and cooling rates in the Venusian upper atmosphere. Separate curves are shown for heating by ultraviolet (UV), near-infrared (NIR) and far-infrared (15 μm) absorptions, and the scales for heating and cooling are different. The challenges involved in producing a complete energy balance model, with dynamics and chemistry included as well, are evident.

gradient. Something like this is clearly happening to produce the mixing that supresses the temperature contrasts lower down. Attempts have been made to revise the model to match the observations, with suggestions that strong eddies might block the transport of heat from day to night, but the details are not very convincing. There seems to be some important physics that we don't understand in the highest region of the atmosphere.

The 'homopause' is the level at the altitude of about 135 kilometres above which the atmospheric gases and their dissociation products start to separate out under gravity. The lightest ones tend to predominate, as the density is so low that there are no longer enough collisions between molecules, atoms and ions to support efficient mixing. Under such conditions, radiative transfer follows different rules, and even temperature as it is usually defined (in terms of the kinetic energy of molecules) has no real meaning.

While the *Pioneer Venus* orbiter swooped in to the thermosphere to measure drag, its neutral mass spectrometer observed the composition, finding CO_2, CO, O, N_2, N, NO, He and H. The atomic oxygen is produced in large quantities in the upper atmosphere by the photodissociation of carbon dioxide by solar ultraviolet and as a result it is the dominant species at altitudes above 170 kilometres, especially during the daytime, followed by atomic hydrogen, helium and molecular hydrogen (Figure 10.9).

Hydrogen is light enough to escape in large amounts by virtue of its thermal energy alone, but at 16 times heavier, oxygen is not, despite plenty of it being available at the upper fringes of the atmosphere. The difference should be enormous, a factor of more than a trillion, and it used to be thought that oxygen could only be lost by being carried along by collisions with escaping hydrogen and helium, a fairly inefficient process. However, recent spacecraft data (see Plate 15) show that water is still escaping from Venus as hydrogen and oxygen ions in roughly the proportion of two to one.

The reason must be that thermal escape is being dwarfed by other factors, the most likely of which is the effect of ablation by the solar wind. This wind is a tenuous medium of

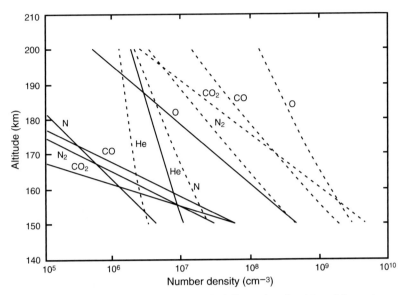

Figure 10.9 Day (dashed) and night-time (solid) number densities of the major gases in the thermosphere from measurements by the mass spectrometer on *Pioneer Venus*.

ions (mostly protons, H^+) and electrons flowing from the Sun and ploughing into Venus's atmosphere at high speed, about 400 kilometres per second. The resulting impacts remove light and heavier atoms, apparently with roughly equal efficiency. On Earth, the solar wind has to interact with the magnetic field, leading to a quite different set of processes and outcomes.

The time for the current atmosphere to lose all of its water at the observed rate is less than 100 million years, which is geologically quite short, raising the fresh question of why Venus has any water at all now. The answer must be that there is a continuing supply, either by comets or by outgassing from the interior. Measurements of isotopic ratios can help to distinguish the two kinds of source, leading to models that estimate the balance between these and all the other climate-modifying processes. The role of volcanoes in resupplying the atmosphere with water is examined in Chapter 11.

Chapter 11
A volcanic world

The key to any discussion of the past and future state of the climate on Venus is an understanding of the production and loss processes for atmospheric gases, from the interior, at the surface, and at the top of the atmosphere where it merges into space. The balance between all of these determines the total mass of the atmosphere, and hence the surface pressure. This, in turn, is the principal factor controlling the temperature and hence the habitability and other characteristics of the surface environment.

At the heart of the problem is an almost complete lack of understanding of the various factors, summarised in Figure 11.1, related to volcanic activity on Venus. Volcanic emissions not only contribute to the mass and composition of the atmosphere, they also fuel cloud formation as part of a complicated cycle of atmospheric and surface chemistry involving various sulphur compounds. The surface itself is mainly of volcanic origin, although this leaves plenty of scope for puzzling about its composition and its capacity for absorbing, as well as emitting, atmospheric gases. The crust is dry and thin but evidently supports some huge volcanic mountains despite apparently being too weak to do so without convective upwelling which, if present, should also drive plate tectonics, although the observational evidence for the latter is elusive.

While there has clearly been significant volcanic activity on the surface of Venus in past geological epochs and there is considerable indirect evidence that this may continue at the present time, the current level of activity is unknown and remains controversial. Estimates for the rate of emission of volcanic gases into the atmosphere range from essentially quiescent to very high levels that may be orders of magnitude greater than those currently found on Earth. This large uncertainty makes it difficult to evaluate the effect of volcanoes on the present-day climate on Venus, or to estimate how conditions at the surface might change in the distant future if, or rather when, volcanism subsides.

As we have seen, there are certainly lots of volcanoes on Venus. A recent study of the images from *Magellan*, which cover more than 90 per cent of the surface, estimated that there may be as many as a million if the smallest vents and domes are included along with the giant volcanic mountains. As further evidence of a volcanic past, the surface has clearly been reshaped by lifting and cracking by plumes of molten rock below the surface, cut by rivers and blanketed by vast fields of once-molten lava. What is less clear is the controversy over how much volcanic activity there is now. Are those *ancient* volcanic cones and

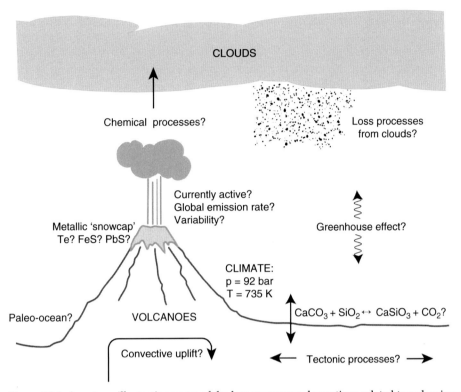

Figure 11.1 A cartoon illustrating some of the key unanswered questions related to volcanism on Venus.

lava fields that we see? Or are they still erupting, filling the air with sulphurous gases and pumping up the atmospheric pressure? Are they directly responsible for maintaining the extreme Venusian climate?

Mars was warm and wet, and had a thick atmosphere, when its volcanoes were active. Maybe Venus is still in this 'pumped-up' state, and will relax to something cooler when the volcanoes finally subside. Or perhaps they stopped emitting long ago, and the pressure and temperature stayed high because there were no oceans to dissolve and mop up the volcanic gases.

Go to any meeting of planetary scientists these days and you will find support for both of these incompatible views, along with the don't knows and the in-betweens. The evidence is incomplete and comes in many different forms. There are the geologists' interpretations of what we see, especially in the detailed radar images from the *Magellan* mission and elsewhere. There are the theoreticians' views of what we expect, based on what we know about the formation of the terrestrial planets and their known properties, the Earth in particular. We can make estimates of the current level of volcanic activity using indirect evidence based on the composition of the atmosphere, which we now know quite well (although not well enough, especially the isotopic and noble gas abundances), the geochemistry of the surface rocks and the flow of heat from the interior of the planet, both locally through 'hot spots', and globally.

Images of volcanic features on the surface

We have no actual pictures of volcanoes on Venus, dead or alive, in conventional images taken with the familiar type of camera or television operating in visible light. There are plans, or more accurately wishes, to one day soar over the giant peaks on Venus with cameras on steerable, floating platforms and settle the argument as to whether they are, even now, belching smoke and lava, or are long dormant. Most likely future robotic explorers will see some of each. They will also sniff the composition of the plumes and see how this compares with earthly volcanic gases and aerosols, and with the Venus atmosphere as a whole. We return in Chapter 17 to discuss the prospects for missions that will accomplish these spectacular but regrettably probably far-off objectives.

Meanwhile, we have the spectacular radar images from the *Magellan* mission. These show massive structures that are obviously volcanoes, even at first glance, since they have vast lava flows that can easily be seen in the radar images from orbit (Figure 11.2). These discharges, possibly in some cases still flowing, often extend for hundreds of kilometres, radiating away from a central craterlike caldera. However, the radar images cannot show the smoke, ash and gas that should be spewing from any of the volcanoes that are currently active. The lava flows often look pristine, but exactly what that means in

Figure 11.2 A *Magellan* radar image of the large volcano Maat Mons, showing the lava flows that extend for several hundred kilometres from the base of the 8-kilometre-high mountain. *Magellan* mapped the volcano from orbit and then the data were computer-rectified to give a perspective effect as if it is being viewed from the surface.

terms of its emplacement could be uncertain by millions of years. The same is true of the low-lying plains that cover most of the planet. It seems clear that these have been filled by lava, which then solidified, but it is hard to say when.

On airless bodies like the Moon, and even on Mars, studies of impact craters provide a valuable measure of the age of the terrain. Models are available of what the flux of impacting objects has been over the age of the Solar System, and this allows the number of craters and their size over a sample of the surface to be converted into its age since it was last covered over. This method does not work on Earth, because on a water planet there is too much weathering and vegetation growth, as a result of which the record is largely destroyed.

Venus is an intermediate case, in that it does have craters that can be clearly identified as the result of impacts (as opposed to being volcanic calderas), but they are relatively few because of the shielding effect of the thick atmosphere, which breaks and burns up the smaller and more frangible meteors before they strike the surface. At the last count, 957 impacts had been identified in the *Magellan* images.

The randomness of the craters on the surface, only a few of which appear to be partly covered over with fresh lava, was interpreted by some experts as indicating that volcanism subsided after a catastrophic, planet-wide resurfacing event roughly 500 million to a billion years ago. They suggest that Venus now has a low level of volcanic activity, and will stay that way until it 'boils over' again at some indeterminate point in the future. This interpretation has been challenged and argued and is still not resolved. We do know that individual volcanoes on Earth behave rather like this, with spectacular eruptions separated by quiescent periods, but it is quite a stretch from that to imagine a million volcanoes all over the planet coordinating their eruptions in the way suggested, and waiting half a billion years between events.

The apparent lack of continental plate tectonics on Venus has relevance for the rate of volcanism, because both are driven by the release of heat from the interior of the planet. On Earth, mantle convection carries heat outwards as it moves continental-sized plates on the surface. If this process is not active on Venus then something like the same amount of heat must escape some other way, which must mean through volcanoes as the only other way is by conduction, and while this undoubtedly occurs, it is far too inefficient.

Although the 'Venus is a one-plate planet' school of thought has prevailed since *Magellan*, and we have pretty much accepted it in the discussions in previous chapters, recent work has suggested this may be too narrow a view. Earthlike plate tectonic activity is detected mainly by observing features in the oceans. The fact that the spreading and sinking zones that are the defining characteristics of continental-scale movements are on the ocean beds and not on the continents, may mean that perhaps we should not expect to find anything similar to Earth's mid-ocean ridges on Venus. Venus does have examples of rift zones, where the surface is torn apart, and mountain belts, where it pushes up, in both cases apparently by relative motions similar to those seen on Earth's land areas. Thus, it is possible that the crust of Venus is globally as well as locally convective, and that the planet has its own version of plate tectonics that we have until now failed to recognise.

Surface composition and temperature

We know that a lot, in fact most if not all, of the surface of Venus is solidified lava, but as well as not knowing when it was emplaced, we do not know its detailed mineralogical composition very well. On Earth, there are lots of different types of lava and Venus is no doubt the same.

It was mentioned in Chapter 9 that the river valley features that extend great distances were probably cut by massive flows of low-viscosity lava that must have had to stay fluid for a long time after leaving the ground. Not only that, but some of the other flows which are wider are so streamlined that the only possible interpretation seems to be that they flowed very fast. In one estimate, the flow speed was as much as 70 metres per second – about 150 miles per hour!

This phenomenal behaviour would not be possible with the kind of lava that is commonly found on Earth unless it had a very high content of water or some other volatile, such as carbon dioxide or certain sulphur compounds. We would love to know what these are, not only to understand the flows themselves and the nature of Venus volcanism better, but also because the volatile material can end up in the atmosphere and affects the climate.

We saw in Chapter 6 that it has recently been possible to make images of the surface at near-infrared 'window' wavelengths. These can be used in several ways to look for active volcanoes. First, since the source of the light making the pictures is thermal emission, hot lava will show up bright against its surroundings. Next, we can try to work out the composition of the lava if a spectrum, at least a crude one, can be measured. In principle, infrared spectra will also tell us the composition of the gases in the plumes, but that needs higher spectral resolution which has not been possible so far.[1]

New data that offer the first two possibilities are available now in the form of infrared emissivity maps of the surface from the *Venus Express* Visible and Imaging Infrared Spectrometer and the Venus Monitoring Camera (VIRTIS and VMC, see Chapter 8). These maps have already been used to search for thermal anomalies that might correspond to volcanic lava fields, recognising that these would have to be very extensive and quite fresh so that they are still much hotter than their surroundings. A further complication is that the surface emissivity varies a lot, and these variations are difficult to separate from temperature contrasts. A bright patch might be at the same temperature as its surroundings, but made of material that emits radiation more effectively. One way to be sure this is not the case is if the lava is so hot that the emission would correspond to a non-physical value of the emissivity (more than 100 per cent of a perfectly black body) at the temperature of the surroundings. Alternatively, there are recent reports of preliminary indications of changes in temperature over time as a fresh lava field cools that might become conclusive as more data is analysed.

---

[1] *Venus Express* carried a high-resolution spectrometer intended to address this goal, and many others, but the Planetary Fourier Spectrometer has not delivered any planetary data due to the scan mirror that directs the instrument pointing becoming, and remaining despite strenuous efforts to move it, stuck in the calibration position. A similar instrument worked well on *Mars Express*.

Recently, a flow that is definitely hotter than its surroundings has been tentatively observed in the infrared. However, the technique is in its infancy and even a state-of-the-art device like VIRTIS lacks the resolution and sensitivity to search efficiently. Also, of course, there is the expediency issue – the lava has to be spotted before it cools. This may have been the case in some of the lava flows as seen in thermal microwave images by *Magellan* (Plate 19). The hottest parts of the lava flow are the bright patches superimposed on the dark background of cooled, earlier flows. Some seem to be at least 85 degrees above the temperature of the surrounding plain, in this case Bereghinya Planitia,[2] possibly indicating recent volcanic activity there.

Infrared spectra provide some information about the composition of the lava-like features emanating from volcano-like structures (Plate 10). Although the sort of spectral observation that is possible at present cannot identify the precise mineral compositional of the lava flows, the data can be used to show that it is probably pristine material that has undergone little surface weathering, in contrast with the ancient lava beds that cover most of the planet. With some intelligent guesses of how long such weathering should take, it is possible to estimate that the flows are geologically young, probably less than a quarter of a million years old, indicating that Venus is currently, or at least recently, active.

Volcanic gases in the atmosphere: sulphur dioxide

All of the gases in the terrestrial planet atmospheres were exhaled from the interior as the planet cooled, and they are all present to some degree in the emissions from volcanoes at the present time. The dominant gases in volcanic eruptions on Earth are water vapour, carbon dioxide and sulphur dioxide, in quantities that vary from event to event. Typically, they range over the following values:

Water vapour (H_2O)	40% to 97%
Carbon dioxide (CO_2)	1.5% to 50%
Sulphur dioxide (SO_2)	0.5% to 12%

with smaller, variable amounts of carbon monoxide, nitrogen, hydrogen, helium, argon, neon and other gases also present.

The proportions on Venus are likely to be different, in particular because the much drier crust will mean much less water. A better guess for Venus would be something like nine parts of carbon dioxide to one of sulphur dioxide, with the other gases together contributing less than 1 per cent. Carbon dioxide is of course the main gas in the atmosphere, so the number we want in this case is essentially the total mass of this gas that escapes, so we can estimate whether it is enough to affect the surface pressure significantly in recent times.

Carbon monoxide, as we saw in Chapter 7, is a significant minor constituent in the lower atmosphere of Venus, at around 35 parts per million. Most of this comes from the dissociation of CO_2 in the upper atmosphere, mixed down by the atmospheric circulation near the poles. Nitrogen and the noble gases have no efficient removal processes and so

---

[2] Bereghinya is a Slavic goddess, protecting woods and waters.

tend to build up in the atmosphere over time. The nitrogen, in particular, can affect the climate over the very long term, through its effect on the surface pressure.

Along with water vapour and carbon dioxide, sulphur dioxide is one of the commonest gases exhaled from volcanoes on Earth. Until we have direct measurements, the expectation is that this will be the case on Venus, too; the circumstantial evidence is very strong. The atmosphere on Venus is loaded with sulphur compounds, with sulphur dioxide itself present in quantities thousands of times greater than we find on Earth. The sulphur in the clouds of concentrated sulphuric acid droplets must have come from volcanoes at some point in the planet's history, most likely quite recently in geological terms since most estimates for the lifetime of SO_2 in Venus's atmosphere are quite short.

When the clouds are included, there may be as much as 10 million times as much sulphur in all forms in the Venusian atmosphere than in Earth's. Much of this difference is attributable to the fact that the troposphere on Earth is purged by rain, and the resulting sulphurous solutions can permeate the land and the oceans so that the sulphur remains lost to the atmosphere permanently. However, in the stratosphere and mesosphere, where even on the Earth the air is very dry and there is very limited downward mixing with the wet troposphere, Venusian SO_2 abundances are still 100,000 times greater, at around 1 part per million versus only 10 parts per trillion for the Earth.

Most of the sulphur in the lower atmosphere is unlikely to have been around for very long because it reacts with the common minerals on the ground, particularly calcite (calcium carbonate). A calculation of the rate of reaction of atmospheric sulphur dioxide with a model of the most plausible composition of the surface comes up with a residence time for the latter in the atmosphere of about 2 million years, very short geologically speaking. To replace that mass of SO_2 over the same time period works out to require the emission of about 30 million tons of the gas from volcanoes, an amount that is not too different from the average sulphur dioxide emission rate from volcanoes on the Earth. So this estimate, admittedly laden with approximations and uncertainties (like most of the estimates we look at in this chapter), suggests that Venus has about the same level of volcanic activity as the Earth. This is pleasing since, until proven otherwise, we naïvely tend to expect the planets to be similar geologically.

On the other hand, the high atmospheric SO_2 could conceivably be part of a closed sulphur cycle in the very hot, dry conditions on Venus, with no large sources or sinks. If the sinks for sulphur (presumably at the surface) are slow or negligible, there would be no need for fresh sulphur and other compounds to be expelled continuously from volcanoes. This could be the case if calcite is not in reality common but instead the surface is made of something that reacts only slowly or not at all with SO_2. It could also be the case if the porosity or the mixing rate of the regolith[3] is small, so that even if calcite is abundant, it could became saturated with sulphur before the atmospheric component reached equilibrium. For the calculation in the previous section to be valid, we implicitly assumed that the surface is sufficiently porous, is churned up by the wind or by earthquakes, or

---

[3] The regolith is the loose or lightly consolidated material, soil, sand, rocks and so on, that covers much of the surface on an Earth-like planet such as Venus or Mars.

resurfaced regularly by lava flows, or something, so that a supply of fresh calcite is always available.

A group of Japanese scientists noted that the observed amounts of SO_2 could in fact be in equilibrium with the surface without any volcanism at all if the rock and soil contained a high proportion of pyrites (iron sulphide, FeS_2) instead of calcite. Such measurements as we have, principally from the *Venera* landers, support the expectation that this mineral is probably uncommon, and laboratory studies have shown that it is unstable at the high surface temperature on Venus in any case. However, a lot of other combinations of different minerals and reactions are possible and some sort of a balance without volcanoes cannot be ruled out until we have really seen what is going on chemically at the surface across the different types of terrain on Venus. This is another reason for obtaining and studying samples returned to the Earth.

The water budget

Water vapour is the most abundant gas in many of the volcanic plumes on the Earth. This may not be the case on Venus, because ocean water is not available to be recirculated through the crust into volcanoes as it is on Earth. However, it would be surprising if Venusian volcanoes did not emit some water, even if much less than their terrestrial counterparts. Evidence that they do comes from observations of water vapour in the atmosphere and liquid water as a component (with H_2SO_4) of the cloud droplets.

Since we also know that there is currently a fairly rapid loss of water from the top of the atmosphere, it is probably being replaced from sources inside the planet. We can try to estimate how large the volcanic source is by working out the total budget of water in Venus's atmosphere and assuming that the net source equals the net sink, that is that the water supplied by volcanoes is equal to the loss to space. This in turn assumes that water on Venus is in global dynamic equilibrium, with the environment overall getting neither wetter nor drier over time.

The mixing ratio of water vapour in the lower atmosphere of Venus is about 30 parts per million, 1,000 times less than on Earth. The equivalent amount for the water trapped in the clouds is about a tenth of a part per million, based on models of the clouds like those discussed in Chapter 12, and therefore a negligible part of the total. Thus the total mass of water in Venus's atmosphere is about 10 trillion tons.[4]

Volcanoes as a source of water

If our starting assumption is that water vapour in volcanic emissions is as abundant on Venus as on Earth, then water would be added to the atmosphere at a rate of about 100 million tons per year, also assuming an Earthlike level of volcanic activity. It would take 100,000 years at this rate to double the amount of water in Venus's atmosphere if there

---

[4] 5×10^{20}kg (mass of the atmosphere) $\times 0.00003$ (fractional abundance of water, 30 parts per million) $\approx$
     10^{16}kg $= 10^{13}$ tons.

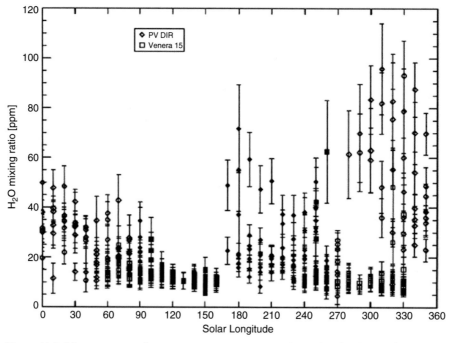

Figure 11.3 Measurements of water vapour concentration above the clouds, as a function of solar longitude (local time of day).

were no losses. On the other hand, if the loss rate was equal to this then the amount would be stable with time, a scenario that always makes us feel comfortable although we cannot be sure it is real.

The average water vapour abundance above the clouds seems to be about one-tenth of that found in the lower atmosphere at between 2 and 60 parts per million. This is not surprising because of the desiccating action of the concentrated sulphuric acid in the clouds; in fact, we might expect the upper-atmosphere values to be even lower. In seeking to explain that, we also have to account for the fact that the abundance varies a lot with time and location (Figure 11.3, see also Figure 1.8), and that it increases with height. *Venus Express* finds that the measured mixing ratio above 70 kilometres reaches 1,000 parts per million, much more than the tropospheric value of around 30 parts per million, although of course the pressure is much lower as well so a smaller number of molecules is involved.

Three different effects are probably responsible, in proportions that are still to be worked out. First, the upper-atmospheric mixing ratio of water vapour might be expected to be anomalously high and highly variable, like that of SO_2, if both gases are being expelled together from large eruptions at the surface. Secondly, the water in the upper atmosphere could, in principle, be supplied by a non-volcanic source, the arrival from space of icy cometary and meteoritic material. Finally, photochemistry and meteorological conditions near the cloud top are mysteriously complicated and are undoubtedly contributing to or even dominating the observed variability.

The current loss rate of water

We can estimate the loss rate due to dissociation of H_2O by solar ultraviolet in the upper atmosphere and subsequent loss of the products by thermal escape and solar wind erosion. A major discovery by *Venus Express* was that there is a large flux of oxygen as well as hydrogen ions in the plasma environment around Venus in which the spacecraft operates. Furthermore, the early indications were that H and O were roughly in the ratio of 2:1, apparently confirming H_2O as the source. We need to integrate these data for all directions around the planet and over time in order to determine the overall loss rate.

The resulting estimate is low, about $2 \times 10^{25} O^+$ ions per second, which corresponds to only about 10,000 tons of water loss each year from the entire planet. Terrestrial volcanoes produce more than 10,000 times this amount, that is about 100 million tons per year. Of course, this assumes that each water molecule lost corresponds to one oxygen ion lost. Until recently, it was thought that only the hydrogen could escape when water was dissociated in the upper atmosphere, as calculations had shown that the oxygen atoms were too heavy. The conventional wisdom was that most of the oxygen stayed on Venus, ultimately ending up in the crust as oxidised minerals.

This probably does happen to some extent, although, like so many things about Venus, the actual quantities and rates are unknown. Removal of oxygen by oxidising various materials in the crust, atmosphere and clouds is plausible but there is no corresponding way to lose hydrogen. If the flux of ions leaving Venus really has only two hydrogen atoms for every one oxygen, it seems that the escape mechanism is more efficient than was thought and all of the water escapes to space as fragments.

Loss of water in the past

The new value for the escape rate is not only thousands of times too small to be consistent with Earthlike emissions from volcanic activity on Venus, it is also more than a million times smaller than the flux of water from the planet calculated by modellers when they estimated that Venus could have lost an Earth-sized ocean in a few tens of millions of years. The latter reflected the belief that water escapes from Venus much more easily than it does from Earth, owing in part to the lack of a planetary magnetic field on Venus. This was expected to allow the solar wind to impinge directly onto the upper atmosphere and remove lightweight species such as hydrogen and helium with high efficiency. Now, however, the *Venus Express* values suggest less than Earthlike escape rates at Venus, despite a radically different near-space environment once thought to favour higher loss rates.

Of course, the present rate could well be much smaller than that which applied billions of years ago when conditions were different. The concentration of water in the dissociation zone in the thermosphere was probably several orders of magnitude larger, increasing the flux of hydrogen by a similarly large factor. The 'blowoff' from this large flux of hydrogen would have made it easier for heavier atoms and ions, including oxygen, to escape then.

Atmospheric erosion by impacts of planetesimals with enough energy to blast gases away into space seems to have been important for the evolution of the Martian atmosphere. The impactors would have had to be correspondingly larger to have had the same effect on

Venus with its higher escape velocity, but there is plenty of evidence on Mercury, Earth and Moon that very large impacts did occur all around the Solar System when it was young.

Finally, the magnetic field and magnetosphere of Venus has probably evolved over time, as has the solar wind, and the rate at which the atmosphere was stripped away by collisions and charge exchange may have been much higher in an earlier epoch.

Balancing the budget

The *Venus Express* estimate of the global water loss rate is preliminary since the spacecraft is still collecting data, and even those that are already in the archive are still waiting for analysis. However, it is the best number we have and must be taken seriously. The most likely possibility that would put the current atmospheric water budget back in balance with the low loss rate is that the Venusian volcanic emissions are not similar in composition to their terrestrial counterparts, but much drier. However, the Venusian volcanic gases would have to have around 10,000 times less water in them than Earth's, even in the case that the total mass of all other gases emitted is roughly the same on the two planets.

The actual composition of volcanic plumes on Venus is unknown, but the percentage of water would be much smaller than on Earth if the Venus crust is desiccated to a great depth, as most geochemists expect. This could be the result of prolonged drying at temperatures of over 450 centigrade, or a consequence of Venus having accreted with much less water than Earth at the higher temperatures prevailing at its distance from the early Sun. Or it could be a combination of the two, which is perhaps the most likely.

The flux of water vapour into the atmosphere required for equilibrium with the escape measurements, as already noted, is of the order of only 10,000 tons per year. It is possible that all of this was originally supplied by an external source, with the volcanic emissions, regardless of the total mass of gas supplied to the atmosphere, contributing a negligible amount of water. The amount that would be required is the equivalent of a single house-sized comet consisting of mostly water ice with a radius of 10 metres or so striking Venus each year, although of course a more steady flow of smaller fragments is more likely, and would not have been detected by any other means. In fact, there is a long-standing controversy about whether there is such a flux into the Earth. One (much-disputed) estimate would have an object of typical mass 20 tons entering Earth's atmosphere every three seconds on average. This would work out at about 10,000 times more than the required flux into Venus to balance the water budget there, so the idea cannot be dismissed.

Noble gases and isotopic ratios

Noble gas abundances and isotopic ratios provide clues to atmospheric evolution because chemical inactivity makes their history easier to trace. For the argon-40 isotope, the only known source is the radioactive decay of potassium-40 in the crust, and subsequent release by outgassing through volcanism. The abundance of argon-40 on Venus is only about a quarter of that on Earth, so the simplest interpretation would be that Venus is no more active than Earth, and possibly somewhat less.

However, the half-life of potassium-40 is more than a billion years and because there is no known sink (except massive impacts, blowing gas off the planet into space, which would have been more important in the earlier times) the radiogenic argon presently in both atmospheres represents emission integrated over most of the history of the planet. Argon measurements cannot therefore tell us much about current levels of volcanism until we can measure the rate at which the concentration is changing at the present time.

Helium is also radiogenically produced in the crust, by the decay of uranium and thorium, and released to the atmosphere through volcanism, but in this case its low molecular mass means there is an efficient sink through its ability to escape to space by thermal as well as by solar wind interaction processes. The mixing ratio in Venus's atmosphere is about 12 parts per million, determined by model extrapolation from measurements above an altitude of 130 kilometres where the ultraviolet spectrum of the molecule can be measured. This abundance is twice that in Earth's atmosphere, while the escape efficiency is about three times lower. This suggests a rate of production that is similar on both planets, averaged over the lifetime for helium in the atmosphere, which is about 600 million years on Venus. Again, therefore, the measurement does little to discriminate current emission from a higher level in the past, perhaps during a major resurfacing episode around half a billion years ago.

Heat flux from the interior

Both Earth and Venus release heat from their deep interiors, via the surface and atmosphere into space. The heat may be residual cooling from primordial times, augmented by radioactive decay and by the conversion of potential energy as heavier elements migrate towards the central core. There is no measurement of how much heat is flowing out of the crust on Venus, but again the expectation is that it should be much the same as the Earth.

We do, however, have reasons to think that the way in which the heat reaches the surface is different on the two planets. Earth seems to be fairly unique among all of the planets in having very active plate tectonics, in which the crust churns constantly and brings heat to the surface. Because it covers the whole surface of the planet, this process is actually more efficient at cooling the interior than the more spectacular, but relatively small and infrequent, volcanic eruptions. If large-scale convection is absent on Venus, the heat flux there may be primarily through volcanoes with only conduction through the crust in addition, and conduction is slow by comparison.

The rate at which the volcanoes would release energy is about 4,000 gigawatts,[5] if Venus has the same internal heat as the Earth. We can get a feeling for this in terms of volcanoes using the mean estimate for the power from Krakatoa during its famous eruption in Indonesia in 1883 (Figure 11.4). As one of the largest and best-known volcanoes, Krakatoa is a convenient yardstick among large volcanic eruptions on Venus as well as Earth. Taking a rounded mean of several estimates, one Krakatoa is about 10 exajoules, or 300 gigawatts averaged over a year.

---

[5] 1 gigawatt (GW) is a billion watts (W), so 4,000 GW = 4×10^{13}W. An exajoule is a billion billion joules (1 EJ = 10^{18}J).

Figure 11.4 Krakatoa, west of Java, in a lithograph from 1888. Earth's most famous volcano, and one of the largest, erupted with devastating force in 1883, and continues to emit gas and dust today.

The eruption of Tambora in 1815, also in Indonesia, was by some estimates even larger than Krakatoa (and more deadly, with the reported loss of in excess of 70,000 lives) but was less well recorded scientifically. The eruption of Mount Pinatubo in May 1991 was the largest volcanic event on Earth in recent times, since the space age began. Meteorologists noted the effect on Earth's climate of all three of these large events, with measurable global cooling (around half a degree in the case of Pinatubo), and increased cloud and rainfall. Tambora probably affected the outcome of the battle of Waterloo.

If Krakatoa released about 300 gigawatts, roughly 100 similar very large eruptions per year would be required on Venus to achieve energy balance by this process alone. This number is uncertain by perhaps as much as an order of magnitude, but still requires a level of volcanic activity on Venus that is substantially greater than that on Earth, or an additional, different, process releasing heat from the interior.

Whether volcanoes dominate, or merely contribute, this model says nothing about whether the heat release is steady or episodic. The release of gases, and therefore heat, from those terrestrial volcanoes that have been studied, has been found to be more the result of steady, low-level emission than from isolated, violent eruptions. For instance, it has been estimated that Mt Pinatubo has added as much gas (and heat) to the atmosphere in the years following its 1991 eruption as it did during the event, one of the largest of the past century. And of course the steady emission is still going on.

Two well-known experts on geothermal activity, Sean Solomon and Jim Head, wrote in 1982, 'The hypothesis that hot spot volcanism dominates lithospheric heat transfer on Venus leads to the prediction that the surface must be covered with numerous active volcanic sources. In particular, if a typical Venus hot spot has a volcanic flux equal to the average flux for the Hawaiian hot spot for the last 40 million years, then 10,000 such hot spots are necessary to remove the Venus internal heat by volcanism.' This works out to a volcanic flux of lava about 300 times that of current Earth, roughly in line with the calculation above.

Bright plumes and ash clouds

Ground-based observers and the camera on *Venus Express* have reported short-lived, localised bright clouds in the upper cloud layer on numerous occasions. One of them can be seen in the image in Figure 11.5. The observers of the largest and brightest event seen recently, which occurred in July 2010 and persisted for about ten days, believed that a large volcanic plume was the most 'attractive and plausible' explanation for the bright features. The eruption would lead to the enhanced formation of sulphuric acid vapour, which condenses to produce a higher, denser, brighter cloud near the top of the normal cloud cover.

The bright, planetary-scale bands and sudden hemispheric brightenings described by the *Venus Express* camera team may also be volcanic in origin. They said that the southern polar region is highly variable and can change dramatically on timescales as short as one day, perhaps arising from the injection of SO_2 into the mesosphere. Observations like these tend to support the idea that major eruptions are fairly common on Venus, although it has not proved possible yet to quantify the emissions from this data.

The bright cloud plumes are composed principally of droplets of sulphuric acid and water that started out as gases in the volcanic ejecta, but if they are anything like the terrestrial version, volcanoes on Venus must also emit enormous amounts of mineral dust and ash. The solid nature of the dust, with relatively large and dense particles, means it does not usually reach the same heights as the hot gas emitted at the same time. Instead, on Earth the dust, entrained with the rising gases, forms a canopy in the upper troposphere, whereas the hot gas is more buoyant and can go on to break through into the stratosphere.

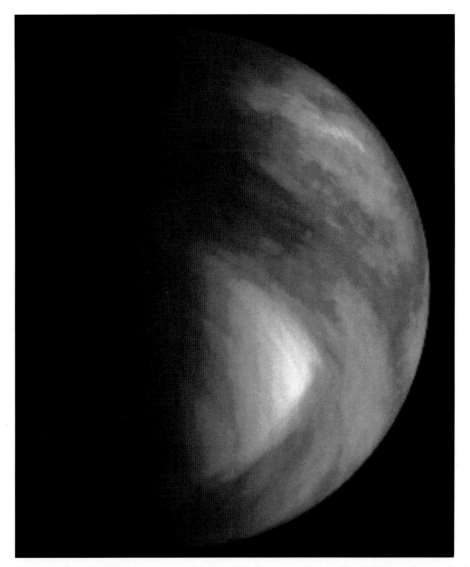

Figure 11.5 A *Venus Express* image of two bright plumes on Venus. The bright patch near the bottom of the picture is part of the bright 'polar hood' which itself is of meteorological origin. The smaller plume near the top of the image looks as if it has been injected from below.

It is interesting to consider whether the higher temperatures and pressures on Venus make volcanic plumes more or less likely to behave like their terrestrial counterparts. A key point is that the major eruptions on Earth reach great heights not through explosive force (except sometimes in the first kilometre or so of their ascent) but by being less dense than the ambient air. The buoyancy is mainly the result of high temperature, but a different composition with lighter gases also helps. Steam is lighter than air on Earth but, as we have seen, there is probably not much steam in Venusian plumes.

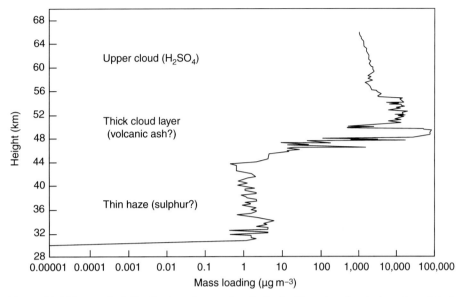

Figure 11.6 The vertical distribution of mass density in the Venusian clouds, based on data from the *Pioneer Venus* probes, shows the hypothetical ash cloud as a narrow layer near an altitude of 48 kilometres. Note that the cloud-density scale is logarithmic, so most of the mass of the total cloud system is in this thin layer.

Carbon dioxide is heavier than nitrogen or oxygen, but not as heavy as sulphur dioxide, which may be the major constituent of Venus emissions. There could also be some water vapour and possibly even significant proportions of extremely light gases such as hydrogen and helium. The temperature of the emitted gas could be higher, relative to ambient, than on Earth. Clearly *in situ* studies are needed before we will know much more about plume dynamics on Venus, but if their behaviour is even remotely similar to those on Earth, we should be able to see ash plumes as dark, absorbing features below the brightly reflecting main cloud layers, using the near-infrared transparency windows to peer through the clouds.

It is possible, even likely, that we already have. Optical instruments on the *Pioneer Venus* large probe found that there is, in some locations at least, a physically quite thin but dense and optically thick, semi-detached layer of cloud at the base of the extended sulphuric acid cloud (Figure 11.6). It was noted at the time that this base layer may have a different composition and origin from the acid cloud; in particular, there is evidence that most of the particles making up this layer are large, and irregular in shape, hence probably composed of some solid material. In a detailed analysis of the particle size spectrometer data, the leader of that team considered the possibility that the large particles in this layer are in fact composed primarily of volcanic ash plumes from erupting volcanoes, and concluded that the existence of large crystalline particles is 'not only possible but probable'.

The analysis of the *Galileo* Near-Infrared Mapping Spectrometer data concluded that the observed variations in opacity in the near infrared are dominated by mode three [large] particles concentrated near the cloud base, and that the results rule out pure sulphur or other substances which absorb less strongly that H_2SO_4 as candidate single-component compositions for these particles. However, the intriguing possibility that mode three particles may be coated particles with a non-absorbing core cannot be ruled out. This last statement might no longer be true because the latest interpretation of *Venus Express* near-infrared spectra, discussed in detail in Chapter 13, finds that the base height for the deepest cloud is at a level that is too hot for sulphuric acid condensate. A refractory material such as volcanic ash would be a more plausible candidate.

Thus relatively large, solid particles might form a large part of the optically thick base cloud detected by the *Pioneer Venus* probes and may make up the deepest, most opaque clouds seen in images in the near-infrared windows obtained from spacecraft and from Earth. At 'window' wavelengths, the clouds are seen as absorbers against the background of thermal emission from the hot lower atmosphere and surface, and hence the darkness of the features corresponds to the total column cloud opacity. If Venus everywhere has a similar structure to that in Figure 11.6, most of that opacity resides in the narrow, dense basal layer of the main cloud at an altitude of about 50 kilometres, which is the densest cloud seen by probes although it is very limited in vertical extent compared with the overlying sulphuric acid cloud layers and the tenuous haze layers below.

The opacity, and hence the density of particles in the cloud, varies dramatically in abundance across the disc of the planet, by a factor of five or more. The patterns often show a ribbonlike structure in near-infrared images, such as those in Figure 11.7 for instance. The ribbons and other, lumpier features might be plumes of volcanic ash, flattening out at the base of the sulphuric acid clouds, possibly because of their encounter with liquid droplets, which would coat the ash particles and tend to arrest their

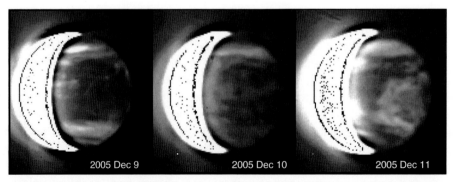

2005 Dec 9   2005 Dec 10   2005 Dec 11

Figure 11.7 Near-infrared (2.3 micron) image of Venus obtained by Jeremy Bailey at the Anglo-Australian Telescope in December 2005, showing dark cloud bands that might be volcanic ash plumes.

ascent in the hot plume. This process would not inhibit the gaseous component which can continue to rise, even in the stratosphere.

The much more tenuous tropospheric haze found below this dense layer in the lower troposphere (Figure 11.6) could be small particles of solid sulphur, or possibly a background aerosol formed of fine volcanic dust from multiple eruptions suspended in the atmosphere. There may also be a link with the ultraviolet markings seen in the upper clouds; for instance, these might be produced by relatively high concentrations of sulphur compounds originating in volcanic plumes. Any more definite interpretation of either of these observed highly space- and time-variable cloud phenomena is difficult without direct sampling of the material and better data on its global distribution and evolution.

On Venus, more so than on the Earth, the density and dryness of the atmosphere below the sulphuric acid cloud base will mean that the mineral dust and ash ejected from volcanoes may stay airborne for a long time before it subsides onto the surface. Thus, it may be that the dust plumes offer the best 'smoking gun' for finding currently active volcanoes, and that we already have copious observations of these plumes in data acquired not only from *Venus Express* but also from ground-based observations made in the 'windows' in the near-infrared spectrum of Venus.

Sulphur dioxide and water vapour fluctuations above the clouds

Previously we looked at the abundance of sulphur dioxide near the surface and discussed how there seems to be too much of it to explain unless there is a steady supply from active volcanoes. The same gas also shows very interesting behaviour high in the atmosphere, above the cloud tops, around 70 kilometres above the surface. Here the abundance of sulphur dioxide is again very much higher than on Earth.

Figure 11.8 shows measurements by the occultation spectrometer on *Venus Express* which, over periods of just days or weeks, found fluctuations at an altitude of 90 kilometres of more than a factor of 100. The pattern seems to be similar at both sunset and sunrise. These are the only times that can be observed by the occultation technique, a factor that is important when considering photochemistry as a possible origin for the variations. Twenty years earlier, using a UV spectrometer that observed sunlight reflected from the clouds all over the dayside, *Pioneer Venus* had witnessed a general decline of nearly as much as *Venus Express*, but over a period of more than a decade (Figure 11.9). These data also showed a lot of short-term variability, superimposed on a long-term decline in the average amount of the gas present.

It is difficult to identify clear patterns from fragmentary data that come from different instruments, sampling not only at different times but also at different heights and locations on the planet. However, it does seem to be clear that there are large fluctuations on both long and short timescales, with a magnitude and speed of variability that is not normally seen in SO_2 or any other minor constituent of the Earth's upper atmosphere – except during a major volcanic event. It is difficult to force the

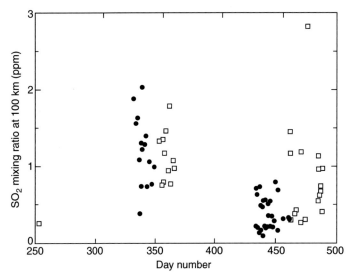

Figure 11.8 Similar to water vapour, large fluctuations are observed in the amounts of sulphur dioxide that spectrometers measure above the cloud tops on Venus. These data from *Venus Express* show the mixing ratio at an altitude of 100 kilometres for sunrise (black dots) and sunset (square dots) occultation measurements over a period of 250 days in 2007.

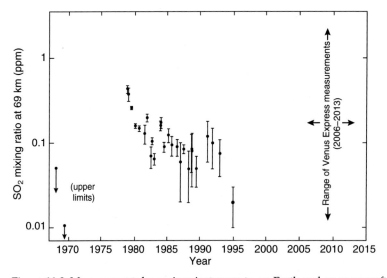

Figure 11.9 Measurements by various instruments, on Earth and on spacecraft, of the amount of sulphur dioxide in the atmosphere above the clouds on Venus, covering a period of nearly 50 years. They show SO_2 varying by a factor of more than 100, apparently with both slow and rapid changes occurring.

changes seen on Venus into any kind of model based on a realistic set of meteorological processes that does not include a variable source. Photochemical production can produce diurnal change, but this does not seem to be the main component of the observed variability in the data or in models. If there is no big source, such as a volcano, some kind of mechanism involving very effective and variable transport must be invoked.

Deep convection is not a serious candidate, since it would be unrealistic to expect much of that to occur at these high altitude levels where radiative energy exchange, rather than vertical transport, governs the transfer of heat. Similarly, eddy diffusion is unlikely to behave in the manner observed for any plausible reason, leaving some kind of large-scale wave or storm activity as the only real possibility. The propagation and breaking of planetary waves has been observed to produce temperature and water vapour abundance variations in the terrestrial mesosphere, although the amplitudes are much smaller than the effect seen on Venus. By far the largest spatial and temporal variability in SO_2 seen in the upper atmosphere of the Earth is that resulting from volcanic plumes, as discussed later in this chapter.

Comparison with terrestrial volcanism

Occasionally, large eruptions on the Earth inject clouds of aerosol and gas to high levels in the atmosphere. As far as visible airborne material is concerned, the main products are clouds of sulphuric acid droplets, which can reach the upper stratosphere, and mineral ash clouds, which generally level out in the upper troposphere.

A quasi-permanent layer of sulphate aerosol, often called the Junge layer,[6] is found at altitudes between 15 and 25 kilometres, or roughly a pressure of one-tenth to one-hundredth of an atmosphere. This is in large part the relict of the sulphur products (sulphur dioxide, carbonyl sulphide and sulphuric acid itself) injected through the tropopause by the larger volcanoes which erupt every decade or so (every century or so for the very largest). There is also a contribution by slower processes including upward diffusion of gases of both natural and anthropogenic origin. The pressure range occupied by the Junge layer on Earth would correspond to altitudes from 65 to 75 kilometres on Venus. A photochemically produced sulphate aerosol layer is also found there; we usually refer to it as the upper cloud deck.

The eruption of Mt Pinatubo in 1991 was monitored from a range of platforms including Earth-observing satellites. Amongst these was NASA's large and well-instrumented Upper Atmosphere Research Satellite (UARS), which was launched on 11 September 1991. The ash cloud from Pinatubo was observed from the satellite to extend up to about 7 kilometres above the surface, while the sulphuric acid aerosol reached to nearly 40 kilometres. This is a pressure level of about 5 millibars, which on Venus occurs near an altitude of 80 kilometres, as the following table shows.

---

[6] Named for Christian Junge (1912–1996), German meteorologist.

Pressure (mb)	Earth altitude (kilometres)	Venus altitude (kilometres)
1000	0	50
100	16	65
10	31	76
1	48	86
0.1	65	95
0.001	80	104

The table compares the heights above the surface at which given pressure levels occur on Earth and Venus. Because there is so much more atmosphere on Venus, the pressure there is quite high at altitudes where it is extremely rarefied on Earth.

Furthermore, the UARS data show that the acid cloud remained fairly compact in latitude and height a year and more after its insertion (Figure 11.10), although not in longitude because the prevailing winds stretched the cloud around the planet in a matter of weeks. The larger, heavier ash particles were removed by dispersal and rainout in the troposphere much more rapidly.

The lower part of Figure 11.10 shows a section through the SO_2 cloud from Pinatubo at a height of 26 kilometres (about 20 millibars, equivalent to 70 kilometres on Venus) as measured by the microwave radiometer on UARS, at a wavelength which is sensitive to this gas but not the aerosol. Again, the distribution is quite compact in latitude, seen here about 100 days after insertion, and this persisted for over a year. If there were multiple eruptions on the scale of Pinatubo at different latitudes during the course of a year on Earth then this would result in the kind of strong inhomogeneity in the global distribution of sulphur dioxide that we seem to be observing on Venus. Furthermore, although the analogy cannot be pushed too far, the pressure levels at which large, volcanically induced SO_2 variations are seen on Earth are not too different from those on Venus, either.

Models for the current level of volcanic activity

Those who believe Venus was resurfaced by a surge of volcanic activity about 500 to 700 million years ago, and has been quiescent since, support a scenario in which Venus has little or no active volcanism at the present time. Great ingenuity has been expended in trying to develop a model for this idea that can be reconciled with the observations. As we have seen, the cratering record on the surface could be consistent with this view, and in fact that is how it came about. What about the other evidence?

A low volcanic activity model for Venus is summarised in Figure 11.11. This assumes the sulphuric acid clouds in Venus's middle atmosphere originate as elemental sulphur in the regolith, which converts into carbonyl sulphide (OCS) in the atmosphere through an equilibrium reaction with the surface. The OCS, with an estimated abundance of 14 parts per billion, is then mixed in the troposphere by the general circulation and some of it raised

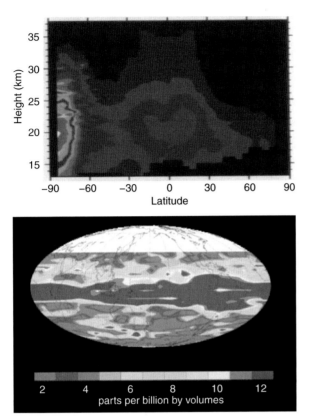

Figure 11.10 Observations from the Upper Atmosphere Research Satellite (UARS), of the vertical and zonally averaged latitudinal distribution of volcanic sulphate aerosol following the eruption of Mount Pinatubo in 1991. (Top) On 21 July 1992, more than a year after the eruption, the Improved Stratospheric and Mesospheric Sounder infrared limb-scanning instrument observed the aerosol opacity at the limb. The large concentration of aerosols over the south pole is due to polar stratospheric clouds (predominantly frozen nitric acid). The sulphate aerosol from Pinatubo is the heart-shaped cloud over the equator, extending up to an altitude of about 40 kilometres. (Below) Sulphur dioxide abundance map of an altitude surface of 26 kilometres for the Earth on 21 September 1991, about 100 days after the Pinatubo eruption, measured by the Microwave Limb Sounder.

to the cloud-top level where it is oxidised to make sulphur dioxide and then sulphur trioxide, following which it reacts with water to produce H_2SO_4, which condenses to form the clouds.

The proponents of this model then postulate that some of the H_2SO_4 propagates upwards to form a layer above 90 kilometres, where it is dissociated to re-form SO_2. The rate of production is sensitive to the temperature, through the reaction rates, and to the supply of sulphuric acid, which depends on the horizontal and, especially, the vertical transport. This relatively dense SO_2 layer in the mesosphere and its hundredfold variability was observed by solar occultation measurements at the terminator, believed to be a region of strong vertical motions. The observed high abundance and variability of SO_2, and the discrepancy with microwave measurements that show much smaller amounts, may be accounted for by a combination of photochemistry and dynamics.

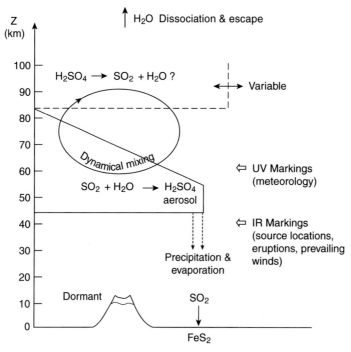

Figure 11.11 This concept for Venus with no significant current volcanic activity assumes that the short-term variability originates in mixing processes in the upper atmosphere that loft sulphuric acid sporadically from the clouds to the 80 kilometres level and higher, where it is dissociated to reform SO_2. The high SO_2 levels near the surface are maintained by crustal materials such as iron sulphide.

The difficulties with this hypothesis include the source at the surface. First of all, it is hard to explain an ample supply of sulphur unless SO_2 is being emitted in large quantities from volcanoes. Then, the estimated abundance of OCS in the near-surface atmosphere is about 10 parts per billion, but the abundance of SO_2 at the cloud tops in the model is set at 100 parts per million. Sulphur conservation would seem to imply that another factor of 10,000 has to be found from somewhere. Finally, the model has to assume unrealistically high values for the vapour pressure of H_2SO_4 and the reaction cross-section for the production of SO_2, compared with currently accepted values, before it works to explain the measurements, which themselves are difficult to reconcile with other observations.

Even if this stretch is allowed and there is sufficient production of SO_2, it is very hard to explain its observed variability by any meteorological process that lofts sulphuric acid from the cloud-top region to 80 kilometres with day-to-day changes of the magnitude observed. Transport from the nightside to the dayside is much easier to accept, and indeed certainly occurs, but again its efficiency is not likely to fluctuate by a factor of 100, and in any case the observed diurnal differences are much too small to account for the fluctuations seen at a fixed time of day.

A competing model in which the mesospheric SO_2 and its variability have their origins in numerous and regular volcanic eruptions at the surface is summarised in Figure 11.12.

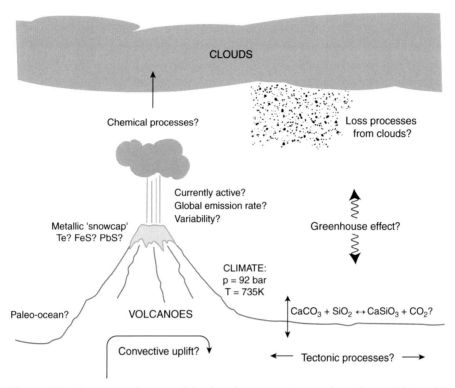

Figure 11.12 An active volcanic model, where large eruptions produce plumes rich in sulphur dioxide which rise buoyantly into the middle atmosphere to produce a layer that is both spatially and temporally variable, continuously depleted by conversion to H_2SO_4. The volcanic ash from the eruptions is trapped at the base of the condensate cloud and forms the patterns seen in near-infrared images. SO_2 is in equilibrium with common minerals on the surface.

A mechanism like this was proposed originally to explain *Pioneer Venus* observations of large-scale changes in the sulphur dioxide concentration and the variable presence of high aerosol hazes above the main cloud tops. Then, however, a single large injection was thought to explain decades of observations, whereas the new data suggest that eruptions large enough to have an impact on the composition of the upper atmosphere must be considerably more frequent. If they are, this can explain the mesospheric SO_2 layer and its variability, the sulphuric acid clouds, the high tropospheric levels of SO_2, and the deep, opaque cloud ribbons containing large, solid particles, all of which are difficult, if not impossible, to represent in the model without volcanism. It is also better, although far from perfect as we have seen, at explaining how Venus might get rid of its internal heat if there is little or no continental plate tectonics activity.

The major difficulty which remains is identifying the mechanism for the transfer of volcanic gases from the dense surface environment all the way into the upper atmosphere. Generally, the larger the eruption, the higher the plume is able to penetrate into the atmosphere. This does not have much to do with the force of the explosion; the jet effect of the blast is dissipated over a relatively short vertical distance, just a few kilometres. Rather, it is the size of the parcel of the hot volcanic gases that is released in a short time. This is (except for the solid ash component) lighter than its surroundings, and can rise a long way because of its buoyancy before it mixes with the ambient air.

An eruption comparable to the very largest known to have occurred on Earth, albeit infrequently, would be able to convey volcanic gases as high as the cloud-top region, where their arrival could explain the large secular changes seen there by *Pioneer Venus*. In one theoretical study of plume dynamics, a dependence of maximum plume height on vent radius was such that reaching an altitude of 80 kilometres would require an erupting vent nearly a kilometre across, again a formidable volcanic event, although there are many volcanic features on Venus with calderas much wider than this (and that of recently erupting Pinatubo, which is not the largest volcano on Earth, is about 2.5 kilometres wide).

These calculations are very sensitive to the assumed parameters, including the atmospheric temperature profile and the likely composition of the volcanic exhalations on Venus, as discussed earlier. In the models the gas is ejected at 270 metres per second and a temperature of a thousand degrees; the composition includes 5 per cent of water vapour as the main volatile. Terrestrial volcanoes are water-driven, and often more than half the gas emitted is water vapour, which is not likely to be the case on Venus where the crust is very dry and there is no recycling of ocean water.

The volatile driving eruptions on Venus may in fact be sulphur dioxide, hydrogen sulphide or some other gas. Any significant proportion of a non-condensable gas lighter than CO_2 will make an important difference to the height reached, as will the temperature of the emitted gas. If the volcanic gas on Venus comes from deep fissures or magmatic plumes, rather than the relatively shallow recirculation of tectonic plates, lubricated by water, as on Earth, then it may have a very different composition and temperature to terrestrial volcanoes. As well as sulphur compounds, there may be proportions of hydrogen chloride and fluoride for instance, and possibly even of hydrogen and helium as well; all of these species are present in detectable amounts in the atmosphere.

Consolidating the models

We can see that both of the competing scenarios that attempt to describe volcanic Venus have their difficulties. If there is little or no volcanism, there should not be so much sulphur in the atmosphere. If there is massive volcanism, there should be a lot more water. We can explain the latter since the emissions must be much drier than Earth's, but then the water that drives eruptions on Earth is lacking in the plume. Water vapour is one of the lightest atmospheric gases and if on Venus its role is replaced by the heavier SO_2, it is harder to conceptualise the plumes rising into the upper atmosphere to explain the observed variations.

The on–off volcanic model, with Venus currently quiescent, needs strange surface and interior properties, and unlikely dynamical behaviour. The notion that the pressure to erupt is constrained and then suddenly violent, then quiescent again, is of course common to individual volcanoes on Earth, but the idea that the whole planet Venus behaves like that in a coordinated fashion is hard to swallow.

The SO$_2$ observations showing a high abundance above the clouds might be explained by the equilibrium between sulphuric acid and SO$_2$, with the former predominating in the cloud region (and condensing to form the cloud droplets), while the latter is the preferred state at higher altitudes. The enormous variability in the amount of SO$_2$ – a factor of a thousand or more – *might*, as some assert, be the result of variations in vertical transport, either by bulk motions or by eddy mixing. This last detail is probably the weakest part of the quiescent volcanic model, since it is hard to visualise meteorological behaviour that would change the SO$_2$ abundance by factors of more than 100 and back again in a few days.

The problems are minimised if we combine the two models. The SO$_2$ abundance and its variability would no longer be such a problem to explain in a combined scenario where volcanic plumes change the composition below the tropopause and dynamical mixing only has to raise the enhanced or depleted air up into the mesosphere.

An 'intermediate' model may also be the answer to the conflict between the different estimates of current volcanic activity on Venus. Some studies of plate tectonics have come to the conclusion that the widely accepted paradigm of Venus as a 'one plate' planet may be too simplistic. Rather than building up heat internally for half a billion years and then exploding, then repeating this cycle, the planet may have made a one-time transition from an interior ventilated mainly by pipe volcanism, with correspondingly massive outpourings of lava to form the plains, to a convective regime in which plate tectonics does in fact occur.

The confusion over this has arisen because, even on Earth, the boundaries of plates are not easy to identify on land masses, but are found in the oceans. On Venus, with no surface water, the high lands and the chaotic lowland terrain could correspond to extended regions of upthrust and subduction that have not been recognised as such because they are masked by other behaviour.

Such a theory fits the data we have better than either of the two 'extreme' models we have discussed in this chapter. The transition from high to lower rates of volcanism would be placed around 700 million years ago to match the analysis of the cratering record on Venus. The flooded plains, rivers and other lava-intensive features belong to the earlier era when most of the internal heat escaped with these flows. Now that convection has taken over the transportation of heat, big eruptions are fewer, perhaps more like current Earth. There are still enough of them to produce the observed cloud plumes and large, long-lived increases in SO$_2$ above the cloud tops; the short-term fluctuations at greater heights are indeed the result of meteorology coupled with photochemistry. There is enough venting of various kinds to produce the cloud streamers in the lower ash cloud, and possibly the ultraviolet markings also, arising from variability in the supply of sulphur, also distributed by winds and waves.

Whether this is the right interpretation or not will require further investigations of the kind we discuss below in Part III. In the meantime, having a model that accommodates at least most of the known facts about the planet gives us a basis for planning future missions to address the controversies. It also provides a platform for extrapolating models of the climate on Venus into the past and into the future, as we will attempt to do in Chapter 14.

Chapter 12
The mysterious clouds

Centuries of speculation about the nature of the clouds on Venus, and many theories and guesses about their composition, some quite exotic, finally led to the definite identification of concentrated sulphuric acid in the 1970s. The clouds were considered likely to have a quite simple vertical and global structure because they looked so featureless through a telescope. Now that they have been studied and analysed, from Earth, from orbit, and by probes parachuting through them, are they still 'mysterious'?

Yes, for a number of reasons. Consider some of the problems we are still pondering:

(1) The clouds form distinct layers. What produces this vertical structure? How much does it vary, across the planet and in time, and what produces the variations?

(2) Only the higher layers are 'definitely' composed of sulphuric acid. The clouds form layers over a range of depths in the atmosphere, and there is evidence that the deeper layers have some different composition, possibly solid particles.

(3) Even the higher layers have impurities mixed in them that we have not yet identified, including the absorber that gives rise to the dark ultraviolet markings. Some sort of variable chemistry is going on, producing localised concentrations and patterns of absorber.

(4) The horizontal structure in the upper cloud layer is produced by waves and by weather that can be at least partially understood through its resemblance to terrestrial meteorology. But the deep cloud also shows dramatic, variable contrasts, and these have a different character. Before these were discovered, they would not have been expected at all, let alone on such a dramatic scale, and the mechanism responsible is still unclear. In Chapter 11 we suggested that they could be dust plumes from erupting volcanoes, but this is not yet confirmed.

(5) As on Earth, the clouds on Venus are a central part of a hydrological cycle of evaporation, condensation and precipitation that is at the heart of a highly interactive climate system. Any variation in cloud density or distribution is going to be linked to changes in the energy balance in the atmosphere, and any long-term changes will affect the atmospheric 'greenhouse' and the surface temperature. We need to understand this in detail.

There is still a lot to do then. Let's see how these puzzles fit with what we already know.

Vertical structure of the clouds

Most of our detailed knowledge of the cloud properties comes from optical measurements. Telescopes on the Earth can measure how the scattered light from the clouds varies as a function of the phase angle (i.e. the angle between the Sun, the planet and the observer) and how this changes as a function of wavelength. The variations turn out to be quite distinctive. The brightness and the polarisation are both sensitive to the particle size, its shape and its composition, and an analysis of them was responsible for confirming concentrated sulphuric acid as the main constituent in the 1970s.

Figure 12.1 shows how the polarisation data discriminate, remarkably precisely, in favour of a particle size with a tight distribution of radii around a mean of 1.05 microns. Exactly what it is about the conditions on Venus that favours this particular size is still to be learned. Other plots, not shown in Figure 12.1, varied the refractive index, which is different for different substances, and showed that a good fit was possible for sulphuric acid but not pure water or any other plausible material. Repeating the analysis at different wavelengths further strengthens the conclusions.

Since then, particle concentration and size measurements have been made from entry probes as they descended through the clouds by parachute. These gather cloud information by looking at the attenuation of sunlight, by shining lasers from the spacecraft into the surrounding clouds and observing the backscatter (Figure 12.2), and by using a small

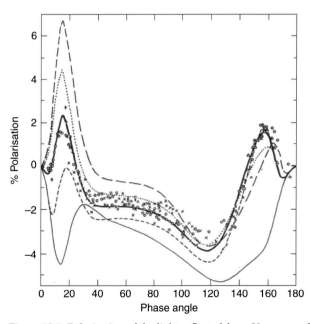

Figure 12.1 Polarisation of the light reflected from Venus as a function of scattering angle. The points are measurements and the curves are theoretical calculations for a model cloud with a range of particle radii *a* from 0.6 to 1.5 microns.

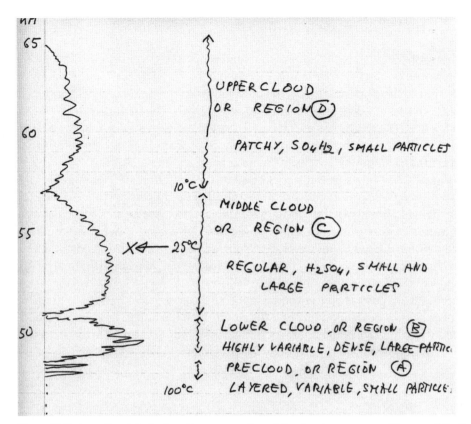

Figure 12.2 A profile of the backscatter cross-section of the clouds as measured by one of the Soviet entry probes, as sketched and annotated by one of the experimenters.

microscope to measure the shadows cast by the droplets on a detector array to determine their size. Spectrometers and polarimeters on orbiting spacecraft operating at visible, ultraviolet and infrared wavelengths have used remote sensing to map the variability of the clouds in time and across the entire globe of Venus by their effect on the backscattered and thermally emitted radiation.

The models pieced together from these data indicate that the clouds cover the planet at heights between about 47 and 70 kilometres, and that they have a complex layered structure. We can think in terms of a representative global mean vertical profile first, in which the gross features of the small number of measured profiles from the instruments on the *Venera* and *Pioneer* entry probes are considered typical of the whole planet. On this are superimposed large fluctuations with space and time in particle concentration and size.

Figure 12.2 shows the brightness of backscattering inside the clouds as measured by a nephelometer instrument on one of the *Venera* probes, sketched by one of the experimenters with his interpretations annotated. All of the probes so far have found that the main cloud consists of three fairly distinct layers, possibly separated by narrow, relatively clear regions with diffuse haze layers above and beneath. This suggests that, whatever variability exists within layers, the general three-layered structure exists over most of the planet, except possibly in the polar regions.

The lower layer (below an altitude of 50 kilometres) is the thinnest in vertical extent, but has the densest concentration of particles, and if we were flying through it in an airoplane it would look the most opaque. It also has the largest particles by quite a large margin, some of them more than 30 microns in mean radius. It contributes most of the opacity in a vertical column from the surface up to space. When we look at the disc of Venus in one of the infrared 'windows', as in Figure 11.7 and Plate 7b for instance, what we are seeing are variations in the particle density, mostly in this cloud layer, acting as a kind of mask attenuating the heat radiation emitted from the hot surface and the near-surface atmosphere.

The big difference in brightness between dark and bright regions shows that this low, physically thin but optically dense layer of cloud has a lot of structure. If we were on the surface looking up, it would sometimes allow us to see the disc of the Sun through hazy thin cloud, and sometimes block our view of the Sun altogether, as cloud can do on Earth.

The upper layer, in the convectively stable stratosphere, is much more tenuous but extends over a larger vertical range. From a base at an altitude of about 57 kilometres, it falls off in density gradually with height over a scale of several kilometres, and is essentially gone by 70 kilometres. This is the layer that is probed by the polarisation measurements from Earth. A very fine haze layer of very small particles exists above the cloud tops and extends up to an altitude of 90 or even 100 kilometres.

The middle layer, from about 49 to 57 kilometres, seems to be more constant in density and is probably quite well mixed by turbulence and vertical motions in the troposphere. The level dividing the middle and upper layers is near the tropopause, where most of the large-scale convective overturning of the atmosphere ceases.

It barely shows in the profile in Figure 12.2, but some of the probes have found a very tenuous layer down below the three main cloud decks, with a physical thickness that is, in places at least, more than 10 kilometres deep. Despite this depth, the particles are so small and so dispersed that the effect on the range of visibility for a camera or astronaut hovering inside this layer would probably be slight.

Particle sizes and composition

The *Pioneer Venus* Cloud Particle Size Spectrometer data found that there seem to be basically three, or possibly four, populations of cloud particles characterised by their average radius. Later studies, most recently with spectroscopic data from *Venus Express*, remain consistent with this basic picture. Figure 12.3 shows how the vertical distribution of the different-sized modes was modelled by *Venus Express* scientists. Because they were using remote-sensing data, obtained looking down from orbit, the detailed vertical structure seen in instantaneous data from an entry probe such as that in

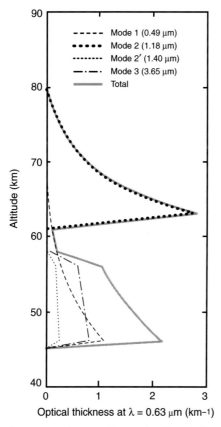

Figure 12.3 A recent Venus cloud model based on an attempt to fit all of the available data, showing the optical depth (at the mean visible wavelength of 0.63 microns) versus altitude for each of the four size distributions or 'modes' with the effective radii shown.

Figure 12.2 is smeared out, and of course those finer details are typical of only one time and place in any case.

Our accumulated knowledge from all the different types of measurement carried out so far indicates that the particles in Venus's clouds range in diameter from a few tenths to about 35 microns.[1] The upper haze layer is composed primarily of the smallest particles, dubbed 'mode 1' by the cloud physicists. The lower haze also seems to be composed primarily of these or similar particles, so they may extend right over the entire range of the cloudy part of the atmosphere.

The upper of the three main cloud layers consists primarily of a second particle type, called mode 2, which has an equivalent radius of about 1 micron, but also includes significant numbers of mode 1 particles. The very abundant, intermediate-sized mode 2

---

[1] On Earth, a typical water cloud droplet is about 10 microns while a raindrop is about a millimetre in radius. Cloud droplets nearly always form around solid nuclei, tiny specks of dust, salt or other materials, and these typically have radii of about one-tenth of a micron.

droplets are mainly responsible for the visible appearance of the planet from outside Venus, and it was these that the early Earth-based observers were measuring when they came up with the first particle size estimates and identified the composition. The mode 2 particles are definitely spherical, and therefore composed of liquid. They can be plausibly explained as being the mode 1 nuclei coated with liquid sulphuric acid solution, which condenses at the temperatures found above an altitude of about 47 kilometres.

A third particle type, mode 3, dominates the middle and lower clouds. Mode 3 particles are relatively large, about 4 microns in radius on average and with some particles as large as 35 micron equivalent radius.[2] The smaller modes are found mixed in with mode 3 at all levels where the latter occurs. Most of the mass of the clouds is in the large mode 3 particles, for which there is some evidence from the inconsistent way they scattered the light beams from the probe for a non-spherical shape. This implies a solid material, perhaps crystals of volcanic dust and ash raised during eruptions and carried by the wind, although the data cannot in this case identify their composition.

A plausible case could be made that the three modes are actually the result of three different particle compositions, as the schematic representation in Figure 12.4 shows. The

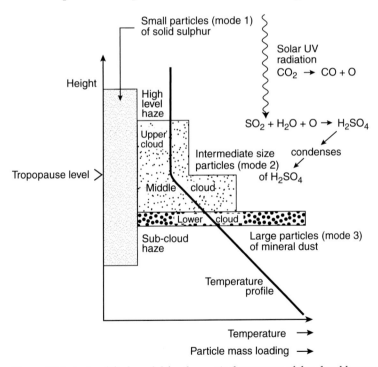

Figure 12.4 A simplified model for the vertical structure of the cloud layers on Venus, typical of most of the global cloud cover but excluding the polar regions. Most of the opacity, and most of the global variability, is in the lower layer, which is physically thin but contains the largest radii and the highest concentration of particles.

---

[2] 'Equivalent' radius means the radius of a sphere that has the same volume as the particle under consideration, a useful definition if the particle is non-spherical.

composition of the smallest particles, which form an aerosol haze extending throughout the cloud layers, whatever it is, probably performs the function of providing the condensation nuclei needed for the efficient formation of cloud droplets. We can speculate that these very small mode 1 particles might be airborne particles of solid sulphur, possibly formed when liquid sulphuric acid evaporates at the temperatures found in the sub-cloud haze. Or they may be some other condensate of volatile substances baked out of the crust at the very high surface temperatures. X-ray spectra obtained from the *Venera* probes suggested to the Russian scientists that significant amounts of phosphorus, iron and chlorine were present in the clouds. These might be part of a cloud of fine dust, raised from the surface by winds or smoke, as distinct from larger ash particles effusing from volcanic vents at the surface.

Liquid cloud materials are easier to identify because they form spherical drops, and a whole range of spectroscopic measurements have confirmed that hydrated sulphuric acid ($H_2SO_4.H_2O$) is the main constituent of the uppermost two of the main cloud layers, and may also form a coating on the large particles in the third, lower layer. The data on layer and particle size can be compared with the results from microphysical models that calculate the cloud properties expected from the composition and the physical conditions in the atmosphere. The models assume the availability of a specified amount of sulphuric acid vapour and calculate the nucleation and growth, and the loss by evaporation and sedimentation, of particles. As might be expected, the models give quite a good representation of the middle cloud layer, which we would expect to be defined by these processes. However, the upper layer is absent altogether until the photochemical production of sulphuric acid from SO_2 is incorporated, and the model underestimates the mass and opacity of the dense lower layer. Again, this suggests that the latter is augmented by some other process not represented in the model, specifically the injection from below of volcanic ash.

Cloud chemistry

It is not too difficult to infer which chemical reactions are responsible for the formation of the sulphuric acid in the cloud layers (Figure 12.5). We have seen that Venus's atmosphere contains large quantities of sulphur dioxide, emitted by volcanoes on the surface, and also contains moderate concentrations of water vapour, probably also exhaled from volcanoes. The two readily combine to form sulphurous acid, H_2SO_3, and in the upper atmosphere the availability of atomic oxygen allows the formation of the much more stable sulphate acid, H_2SO_4.

The supply of atomic oxygen comes mainly from the dissociation of water and carbon dioxide by solar ultraviolet radiation, which penetrates the atmosphere only down to the cloud-top region. The atomic oxygen combines with SO_2 to produce the trioxide, which reacts rapidly with water vapour to make H_2SO_4. Initially, the sulphuric acid forms as vapour, but at the low temperatures in the upper atmosphere (about 35 below zero centigrade at the cloud tops) and in the presence of condensation nuclei (the mode 1 particles), it condenses and forms liquid droplets.

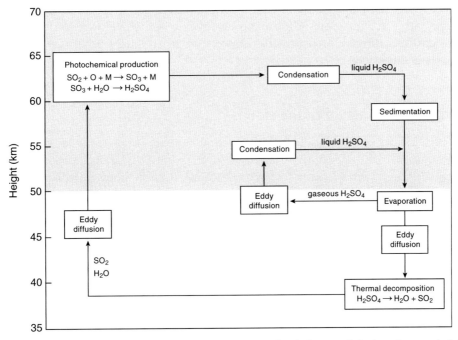

Figure 12.5 A schematic diagram to illustrate how the clouds form and dissipate by coupled dynamics and photochemistry.

At the level where these photochemical reactions are possible, where the overlying atmosphere is optically thin and short-wave solar radiation can penetrate, there is little dynamical uplift and the droplets will tend to sink. The remaining acid vapour will also tend to diffuse downwards to regions of lower concentration. Once inside the troposphere, the convective upwelling forms a thicker cloud by supporting higher concentrations of particles and drawing up a supply of vapour from below, including recycling that produced by the evaporation of large droplets that rain out into the hot lower atmosphere. The result is the denser and more extensive 'middle cloud' region.

The lower cloud is the most opaque, despite being relatively thin in the vertical dimension, but is also the hardest to study, as there are very few measurements. If it does contain large, solid crystalline particles, the most likely origin for these is mineral dust, which may be raised in the plumes from erupting volcanoes. In Chapter 11 we argued that the fan- and ribbonlike structure of the clouds supported this interpretation. An alternative that has been suggested is wind-blown dust from the weathering of the surface, although it seems unlikely that this would reach these high levels as the winds are quite low in the dense atmosphere near the surface. Another explanation might be crystals of some refractory material formed chemically in the hot atmosphere and suspended as a kind of Venusian heat haze, although again this is hard to pin down. They might be large aggregates of the sulphur particles similar to those hypothesised to form the subcloud haze: it is possible to imagine some kind of efficient coagulation process going on near the level where the temperature is high enough that the liquid sulphuric acid droplets are evaporating at a high rate. A future probe, floating inside the lower cloud and analysing the particles directly, will provide answers: at the moment what we have is largely informed guesswork.

Horizontal structure of the clouds

With a picture of the overall vertical structure in mind, we can now look more closely at the question of horizontal variability. Contrary to early indications from the bland appearance of Venus as seen through telescopes, we now know that the variations with location and time are large and actually quite dramatic. Much of the recent progress in Venus cloud studies is in observing and understanding global patterns and trends, especially from equator to pole, with a big boost recently from more than five years' scrutiny from orbit by *Venus Express*.

These data, especially the infrared spectra and images from VIRTIS, an example of which appears in Figure 12.6, can be analysed to obtain information about the cloud thickness as a function of latitude and longitude at various wavelengths, and from this the changes in cloud-base altitude, sulphuric acid concentration and particle size can be deduced. The picture that emerges is of short-term, quasi-random changes, systematic variations across the planet and possibly also some long-term trends.

Looking first at the random changes, the more compact patterns formed by regions of thick and relatively thin clouds suggest large-scale cumulus dynamics, from the same

Figure 12.6 A *Venus Express* image of details in the cloud patterns on Venus obtained in the 2.3-micron near-infrared window. Here the contrast is reversed so that the most absorbing clouds appear bright, to match what we expect to see when we look at them at visible wavelengths.

family of phenomena as thunderstorms on Earth. The cloud material actively condenses and dissipates in rising and falling air associated with energetic weather systems, forming clumps of dense cloud separated by clearer spaces. This is apparent in the 2.3-micron image of Venus from *Galileo*-NIMS shown in Plate 7.

We have also discussed the possibility that these and other, more linear, features may be associated with dust plumes from volcanoes; if this is the case then the irregular coverage of the clouds is due more to the distribution of active sources than to meteorological processes. Transport by winds will still be important in determining the appearance of the cloud, and this is a function of the height reached by the plume. It seems likely that several different processes are going on – eruptions, convection, transport – that we cannot separate without much more data, including some measured *in situ*.

The possibility that the lower cloud deck contains volcanic ash is supported by some recent radiative transfer modelling which shows that the brightness variations seen from space can be accounted for by varying the optical depth only in the lowest few kilometres of the cloud. Another finding, that the cloud base is higher in regions of low optical depth, also supports this hypothesis. Raising the cloud-base altitude in the model by entirely removing the cloud from the lowest 4 kilometres cuts the total cloud optical depth in half, which could account for the bright regions between dark clouds in the 2.3-micron images. In the close-up example from *Venus Express* shown in Figure 12.6, the contrast in the image has been reversed, so the region where the lower cloud seems to be thin or absent is the dark band down the middle of the image.

The systematic variations in cloud over a hemisphere from a recent study are summarised in Figure 12.7. We would expect the base of the sulphuric acid cloud to occur at the height corresponding to the temperature where the acid vaporises, so the altitude of the cloud is expected to depend on the temperature, the acid concentration in the cloud droplets and the rate of supply of condensable vapour. It does in fact appear to fall sharply near the pole, where the measured temperature within the cloud deck is typically 30 degrees cooler than it is at the equator.

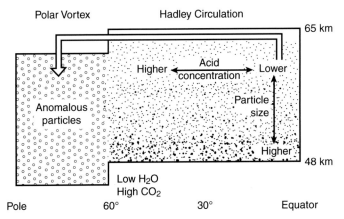

Figure 12.7 A model of the latitudinal variation in mean cloud properties, deduced from radiative transfer models fitted to *Venus Express* spectroscopy data.

At the same time, the average acid concentration in the cloud droplets increases with latitude to a maximum value in the polar collar at about 60 degrees latitude, and also shows a correlation with cloud optical thickness. The lowest acid concentration found is 70 per cent in the very brightest equatorial regions, increasing to 87 per cent in the darkest regions in the collar. This is somewhat counter-intuitive, because the photochemical production rate for the acid is greatest near the cloud tops at the equator.

The acid concentration should also depend on the amount of water vapour available, of course, and the water variations in the cloud are difficult to measure. The explanation is probably to be found in the transport processes that distribute the water and acid vapour vertically and horizontally. The Hadley circulation will deliver high concentrations of acid from low to high latitudes at the cloud tops, and there is some evidence that it also raises water-rich air from the depths near the equator. The combined effect apparently is to concentrate the acid at higher latitudes and dilute it in the tropics.

The new data, at least in this interpretation, also indicate that the average particle size and the water vapour abundance are both higher in bright regions. This is the opposite of what would be found in a convective cloud on Earth, where regions of thick cloud occur in regions of upwelling moist air, whereas cloud is removed in regions of downwelling. Descending air comes from colder regions aloft and is usually drier, plus larger particles are preferentially removed as they fall out faster. On Venus, if the clouds consist of relatively small sulphuric acid solution droplets overlying mode 3 particles that are partially or completely solid, the loss of sulphuric acid droplets as a result of lower condensate amounts or evaporation because of higher temperatures would raise the mean particle size. Sorting out these competing processes is a goal for a future long-term instrumented station floating in the clouds.

Other local particle size variations are more intriguing and even harder to explain. The global observations of Venus in the spectral windows by the near-infrared spectrometer on *Galileo* showed contrasts in the brightness that are different at each of the three wavelengths. Figure 12.8 shows an example of a plot of the radiance at one wavelength against another, in which two branches emerge, indicating two separate sets of cloud properties. If the origin of each branch is mapped onto the nightside of Venus, we see that one branch corresponds mostly to two big patches in the northern hemisphere.

Remarkably, other combinations of wavelengths come up with five distinct branches, suggesting different mixes of the particle size modes. Why should there seem to be five distinct cloud regimes in this way? Either the sources of each type of cloud are different, or there is some process at work that sorts them into preferred size distributions and produces their large-scale spatial correlation.

The models that best fit the data offer a clue. Again they suggest that the variations were largely confined to the lower cloud, and were mainly attributable to varying amounts of the large mode 3 particles. If this is further evidence that the dense cloud features, and the particles in the lowest cloud layer, are the products of massive volcanic plumes, the regions with different particle size distributions could be the products of different major eruptions. The individual events could produce dust with characteristic sizes, and also inject it to different height levels where the winds and fallout rates are different. This is

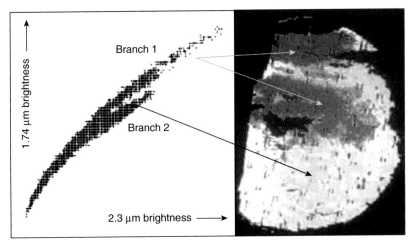

Figure 12.8 Two kinds of cloud: the plot on the left shows how the brightness in the 2.3-micron window varies with the corresponding brightness in the 1.74-micron window. The plot on the right shows how the two branches map onto the nightside of Venus.

unproven, however, and again we have to look forward to an *in situ* explorer to investigate and shed some light.

The ultraviolet absorber

The model of the clouds in Figure 12.4 plausibly describes most of the main features of the clouds but does not include any explanation of the earliest-detected phenomenon linked with horizontal inhomogeneity, namely the ultraviolet markings. While observations of Venus at visible wavelengths, whether made by an observer at a telescope or by a camera on a spacecraft, show little or no detail in what seemed to be a uniform blanket over the whole planet, it has been known since the 1920s that photographs taken through an ultraviolet filter show blotchy and streaky dark features in the cloud. Clearly, some ultraviolet-absorbing substance is non-uniformly dispersed through the clouds, and changing with time, perhaps mainly as a result of mixing from below.

Sulphur dioxide behaves in this way, and is definitely present in spectroscopic observations. This gas absorbs in the ultraviolet, and its spectrum matches that of Venus at some, but not all, wavelengths. Some other material, probably another sulphur compound or perhaps one of the allotropes[3] of elemental sulphur, which also absorbs ultraviolet but not visible radiation, must be contributing as well. Chemical pathways exist that can produce small amounts of elemental sulphur of various kinds as a by-product of the main reactions producing the concentrated sulphuric acid that is the main constituent of the upper cloud.

---

[3] Allotropes are different forms of the same element, characterised by different crystal structures or bonding arrangements giving different physical, chemical and spectroscopic properties. Well-known examples include graphite and diamond (both carbon), and oxygen and ozone. Sulphur has 22 known allotropes.

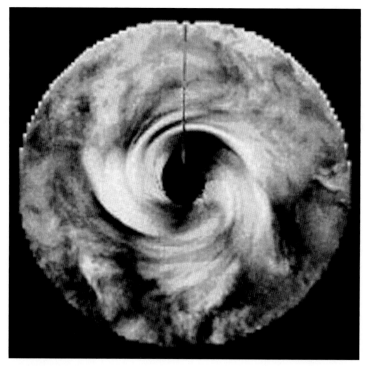

Figure 12.9 A view of the cloud structure surrounding the north pole in a mosaic of images obtained through the ultraviolet filter in the camera on *Mariner 10*.

Whatever their origin, the ultraviolet markings are fascinating because their structure, movements and evolution trace to the cloud structure and meteorological activity on Venus. It was, of course, observing the global-scale 'sideways-Y' feature which rotates around the planet in only four to five days that led to the discovery of wind velocities of 100 metres per second in the cloud tops. The planet rotates at only about 4 metres per second, or once every 243 days.

The ultraviolet markings delineate the polar vortices as well. Figure 12.9 is a view centred on the north pole, made up from several *Mariner 10* pictures (obviously a mosaic since both sides of the pole are illuminated simultaneously, which never happens on Venus) and showing, with the contrast strongly stretched, the influence of the polar vortex on the cloud structure.

Carbon dioxide clouds?

Intriguingly, there is some evidence that clouds of frozen carbon dioxide crystals may sometimes form on Venus, above the sulphuric acid clouds. These 'dry ice' clouds are quite common on Mars, where temperatures are generally cooler, but sufficiently low air temperatures can also occur occasionally on Venus in the minimum in the vertical profile at an altitude of about 75 kilometres, a short distance above the top of the sulphuric acid clouds. The first evidence came in infrared maps from the *Pioneer Venus* orbiter of a cold, high absorber, presumably condensed CO_2, near the tropopause above the dawn terminator, where the atmosphere is coldest as it comes to the end of the long (more than 50 hours at this altitude) Venus night. Since then, similar phenomena have been glimpsed by the infrared instruments on *Venus Express*.

The polar clouds

It is significantly more difficult, in engineering terms, to insert probes or balloons into the atmosphere at the high latitudes where a different cloud structure might be expected in the polar vortices, and so far this has not even been attempted. So, we have no direct, *in situ* measurements of any of the properties of the clouds in the polar vortex region and are limited to remote-sensing data. These show that the vertical structure and the properties of the cloud particles in the vortices are indeed different from those found over the rest of the planet.

We have already seen that the cloud-base altitude apparently decreases sharply with latitude polewards of the collar, reaching at the pole at least 6 kilometres below the equatorial value. The temperature decreases between the equator and pole within the cloud deck such that the isotherms are 4 kilometres lower at the pole than at the equator, and the downwelling that takes place in the vortex will supply additional sulphuric acid and cloud nuclei, reducing the water vapour abundance and probably increasing the coagulation rate. It seems reasonable, therefore, that a decrease in the cloud-base altitude between the equator and pole is compatible with these thermal and dynamical constraints.

However, there remains a qualitative difference between the spectral properties of the cloud particles inside the polar vortex compared with outside that is still not understood. Most of this difference would be explained if the cloud particles are considerably larger at high latitudes. An alternative explanation is that a different chemical species with different absorptive properties to those of concentrated sulphuric acid is present in the polar clouds. Species that have been proposed from time to time as possible cloud constituents include phosphorus anhydride, phosphoric acid and iron chloride. Phosphorus- and iron-bearing rocks are common on the Earth and were reported by the *Vega* probes to be present on Venus, providing a possible source.

Even if they exist, however, it is not easy to come up with potential mechanisms for concentrating these materials into the polar clouds. The possibility that the cloud deck extends deeper at the poles, reaching down to altitudes into the convectively unstable region, might provide a potential transport mechanism for species from the surface to reach cloud altitudes. Also, these particles should be more stable under conditions in which sulphuric acid droplets would evaporate, which could also provide an explanation for the increased ratio of large to small particles.

Diurnal and long-term variation

In general, the variations in the cloud properties as a function of local solar time when measured around the equator are smaller than those that occur as a function of latitude.

This is not surprising, with the rapid zonal rotation of the atmosphere tending to smooth everything out. However, there is a small but significant rising and falling of the cloud-top height by about 2 kilometres during the course of the day that was first observed by *Pioneer Venus* in 1979. The maximum height is in the afternoon, when convective activity because of solar heating is greatest, and a minimum before dawn.

There is some very tentative evidence for a small secular decrease in several of the cloud-related parameters decreased over the first 1,000 orbits of the *Venus Express* spacecraft. For this period, nearly three Earth years, the acid concentration in the cloud droplets, the water vapour abundance and the CO abundance all show a gradual decrease superimposed on global and other shorter-term variations. However, other instruments on the same spacecraft have reported an *increase* in sulphur dioxide over that period, especially a sharp increase in early 2007 compared with previously retrieved values. If the SO_2 abundance variations above the cloud are due to increased volcanism, the cloud concentration decline cannot have been due to a decrease in volcanic activity at the same time, at least not in any straightforward way. Another possibility is variations in the level of solar activity, which may lead to slow changes in the abundance of sulphuric acid. Sunspot activity was decreasing at the end of the most recent solar cycle from early 2006 to late 2008, so potentially this could provide an explanation for lower abundances of photochemical species like H_2SO_4, even if there was an increase in SO_2.

So, all things considered, there are plenty of puzzles still to be resolved before we will really understand the clouds on Venus.

Chapter 13
Superwinds and polar vortices

The circulation and dynamics of Venus's atmosphere behave in ways that sometimes remind us of terrestrial meteorology, but mostly seem quite bizarre. Yet we routinely compute the dynamical behaviour of Earth's atmosphere, for weather forecasting and other reasons, and it should be possible to do the same for Earth's twin. However, even the most basic behaviour on Venus, the four-day 'super-rotation',[1] is proving hard to diagnose or to replicate. Despite a great deal of attention by groups using some of the most sophisticated computers and models, Earthlike simulations with Venusian parameters inserted have tended to circulate too slowly.

A lot of meteorological activity – weather – has been observed in Venus's atmosphere by orbiting spacecraft and measured *in situ* by descent probes, but understanding and interpreting what is going on is still at an early stage. Researchers continue to argue about whether lightning occurs, and although it probably does, there is no clear picture of how or where it is generated. Some of the small-scale and transitory features, such as the waves seen in the ultraviolet images of the cloud tops, and some of the global and semi-permanent behaviour, for example the Hadley circulation, have some recognisable relationship to similar behaviour on the Earth, although they may have to be scrutinised closely to see it. Other important phenomena seem fairly unique to Venus, not just the fast winds that circle the equator, but the complexity of the giant vortices at each pole, and the behaviour in the upper atmosphere, where the circulation seems to change to a completely different regime.

A lot of the difficulty that remains in understanding Venusian meteorology, even after 50 years of exploration by spacecraft, lies in the fact that comprehensive measurements of dynamics are particularly difficult to make. Except for cloud tracking with its obvious limitations, the familiar problems caused by limited data are particularly severe where this topic is concerned. More and better measurements remain the key to progress (Figure 13.1).

Isolated measurements from probes can do a good job of determining the composition of the whole atmosphere in most cases, but provide only very localised information on the

---

[1] The term 'super-rotation' refers to the fact that the winds are in the same direction as the rotation of the solid planet, but much faster, in the case of Venus at the cloud tops about 50 times faster than at the surface below.

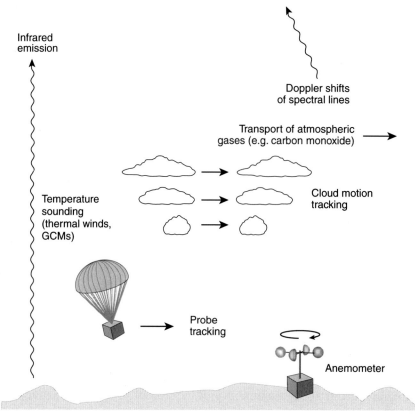

Figure 13.1 A sketch showing the various ways the dynamics of the atmosphere have been measured or inferred.

general circulation. Remote sensing from satellites is great for getting global coverage, but essential detail is always missing, especially in the vertical structure. On Mars, simulations using modified terrestrial global circulation models do a good job of filling in the gaps, but these do not work so well on Venus. This is quite worrying, since these models have the same core as those we rely on for weather and climate predictions on Earth.

Meteorological measurements on Venus

The most basic measurement of atmospheric dynamics was made by the Russian landers *Veneras 9* and *10*. These incorporated the first successful weather stations on the surface, using a simple cup anemometer mounted high on the spacecraft to gauge the wind speed a few metres above the surface of Venus. The anemometer is the rotating vane device that is familiar from most earthbound meteorological stations, spinning on the end of a short mast above a screen containing temperature, pressure and humidity sensors. The winds it found on Venus are light, around 1 metre per second, as we might expect given the high air density and the limited penetration of sunlight down to the surface. A breeze on the Earth would be about 10 metres per second, and the typical wind speed on the surface of Mars is about the same or higher.

Vertical profiles of the wind direction and strength on Venus have been obtained several times by following the drift of descent probes as they pass through the atmosphere. The *Venera* landers were tracked by the measurement of the Doppler shift in the radio signal from the spacecraft. The *Pioneer Venus* probes used an interferometric version of the same technique involving more than one receiving station, which gives better precision in the position and movement of the probe. Because they are suspended on parachutes during most of their descent, they couple well with the wind and this approach gives accurate results, with vertical coverage from an altitude of about 80 kilometres down to the ground. The vertical resolution is excellent, too, showing a lot of detail and resolving regions of high shear (i.e. rapid change in the wind vector). The main limitation is the lack of coverage in space and time; it is very difficult to interpret from the measurements of a single probe the profile of a quantity that varies a lot in time and location. Missions that release salvos of at least 16 (and, if they are small, possibly many more) well-separated but simultaneous probes are an attractive option for a future programme.

Probe wind measurements

The probe results confirmed the very high winds near the cloud tops. Before these *in situ* measurements were made, not everyone believed that the observed movements of the ultraviolet markings in the cloud corresponded to real winds, because the average velocities are so high. At over 100 metres per second, they are stronger than a severe terrestrial hurricane.[2] Such high winds are seen blowing consistently over global-scale distances on the Earth only at very high altitudes in the thermosphere, where densities are very low and friction and turbulence offer slight resistance. On Venus, by contrast, the ultraviolet markings and hence the hurricane-force winds originate at pressures only a little less than those right at the surface of the Earth, where such high winds are never experienced.

The wind profile obtained as each probe continued its journey down to the surface showed that there is a steady decrease with height from the high values in the zonal direction near the cloud tops, down to small speeds matching the anemometer values at the surface. At the same time, of course, they also measure the other component of the wind, that in the meridional direction (equator to pole). This has much lower values, as expected, just a few metres per second maximum, but with a very complicated vertical structure, including alternations in the direction of the wind at heights that were the same, roughly at least, for each of the probes. These could mark the passage of the probe through the different components of not just one but a stack of three Hadley cells, each extending from the equator to high latitudes. The layered eddy sources and sinks which have been postulated as driving the zonal super-rotation may be related to the cell interfaces in this scenario, but this is very speculative.

---

[2] The official definition for a 'strong' hurricane is 50 to 58 metres per second. The worst ever recorded on Earth, easily exceeding the 'devastating' threshold at 70 metres per second, was 89 metres per second or over 200 miles per hour.

Winds are not the only variable that is important for investigating the dynamics of the atmosphere. Most of the probes have confirmed that the temperature gradients in the lower atmosphere are close to adiabatic in the vertical,[3] and close to zero in the horizontal, again as would be expected theoretically from the high opacity and high density. If the vertical rate of change of temperature is higher than the adiabatic value, the lower layer may be less dense and start to rise, falling again when it cools at greater height and thus instigating convective overturning. Conversely, small departures in the measured profiles to values less than the adiabatic temperature gradient show which layers in the troposphere are stable against convection.

Cloud images and infrared remote sounding

The first entry probes confirmed the high wind speeds that had been inferred by tracking the cloud features seen in ultraviolet images from the Earth and from spacecraft, beginning with *Mariner 10*. This opened the door for global wind field data to be obtained. It was only for one level in the atmosphere, but it was the important one at pressures similar to those where the main cloud cover lies on the Earth.

Pioneer Venus, *Venus Express* and other orbiting spacecraft made sequences of images specifically to extract wind velocities over a wide range of latitudes. These showed that the zonal super-rotation is nearly constant from the equator up to the edge of the polar vortex, but then declines steadily to the pole (Figure 13.2). A careful analysis of small-scale features reveals the equator-to-pole velocities as well, nearly 100 times slower.

The velocities derived by this method apply to the region where the pressure is of the order of, or somewhat less than, 1 atmosphere, but the vertical levels being probed are not precisely defined, and may vary from place to place, which complicates any interpretation. The probes, of course, monitor the pressure as they descend, so the winds they measure are accurately tagged with pressure and height information, and we can associate the movements of the cloud markings with the height level at which the probes measured similar speeds. Then we find that the cloud-tracked winds relate to the level of altitude at somewhere between 50 and 60 kilometres.

The data in Figure 13.2 span more than 30 years and significant changes can be seen during that time, although again it is also possible that the markings being tracked occur at changing cloud height levels. Recently, the *Venus Express* camera team reported a 30 per cent increase in average wind speed over a five-year period. While these data are all from the same spacecraft and instrument, it is still possible that at least part of the change is due to variations in the distribution of the mysterious ultraviolet absorber. Such changes could not be too large without being detected by other methods, however, so it seems likely that the super-rotating wind field really is subject to secular variations, perhaps a very slow oscillation with a period of many years.

Cloud-tracked wind measurements are no longer limited to the cloud-top region, since the discovery of the near-infrared windows as related in Chapter 6. Motions in the deeper

---

[3] An adiabatic gradient is one that is just stable against convection, so that when any two adjacent layers are considered, the one on top has a slightly lower density than the one below.

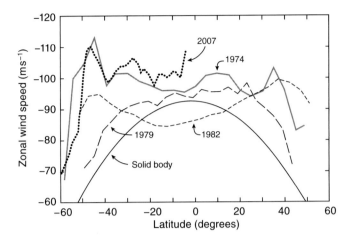

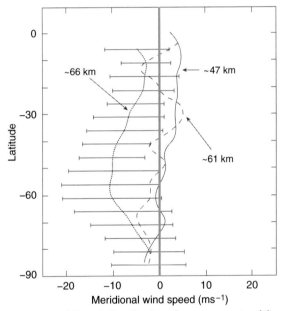

Figure 13.2 (Above). A collection of measurements of the atmospheric super-rotation by cloud tracking in ultraviolet images. The zonal winds are plotted against latitude at various times from *Mariner 10* in 1974 to *Venus Express* in 2007. In addition to the large time variations, note the mid-latitude jets (maxima), and the fact that the atmosphere does not rotate like a solid body. (Below) Equator-to-pole winds measured by *Venus Express*, also using cloud tracking but at different ultraviolet and infrared wavelengths which probe the three approximate height levels indicated. The error bars, for clarity shown only for the data at 66 kilometres, are typical of the other two altitudes as well, and also of the uncertainties (around 10 metres per second) in the zonal wind measurements.

atmosphere were first observed by near-infrared imaging carried out on the nightside of the planet from the *Galileo* flyby in 1990. The features being tracked in the windows originate in the dense main cloud deck near an altitude of 48 kilometres, illuminated from below by infrared radiation from the hot lower atmosphere. The typical velocities inferred near the equator were about half as fast in the zonal direction, and about the same in the meridional, as those from tracking the ultraviolet markings some 20 kilometres higher. Again this matches the height variations of wind and cloud opacity measured directly by the *Pioneer* and *Venera* entry probes.

Temperature and dynamics are linked through various fundamental equations, so an infrared temperature-sounding instrument on a satellite can obtain information about the global dynamics in three dimensions. These data apply to the whole vertical range from which emission is observed, although they still have limited vertical resolution because, even for a single wavelength, the emission originated in a layer typically more than 10 kilometres thick. Measurements of the upwelling flux at different wavelengths can be used to reconstruct temperature profiles first, and then the corresponding wind fields can be computed by feeding the temperatures into a computer model. The method has its limitations, for example because of difficulties with the parameterisation of viscosity, particularly that resulting from eddies, and allowing for the effect of clouds as they absorb and scatter infrared radiation. In principle, the method can be extended further downwards using the near-infrared windows, but a microwave sounder, insensitive to clouds, would be better.[4]

Systematic sounding was carried out by *Pioneer Venus* using five infrared bands at wavelengths near 15 microns to cover the vertical range from 60 to 105 kilometres with a mean vertical resolution of about 10 kilometres. Radiance measurements were made five times per second on a spacecraft spinning at 12 revolutions per minute. The fast sampling used the spin to scan the planet and resolve features a few tens of kilometres across.

The average temperature field obtained in this way for Venus is compared with similar representations of Earth and Mars in Figure 13.3. They all show several features clearly related to the general circulation, including the interesting tendency for the temperature across a broad altitude range to increase from pole to equator, in spite of the fact that the trend in radiative heating is in the opposite direction. This is possible because of dynamical heating as a result of the compression of the polar air mass in the general circulation. On Earth, and especially Mars, polar warming occurs but is more episodic and seasonal; on Venus it is more of a steady phenomenon that is always present.

The Hadley circulation

In terms of pressure and density, Venus's atmosphere consists of a deep, 'oceanlike' region, the middle 'Earthlike' part, and the low-density upper atmosphere, each with a

---

[4] A microwave temperature sounder was originally selected to fly on VOIR, the precursor to the *Magellan* mission, but was dropped during a mass- and cost-cutting exercise part-way through the development of the mission.

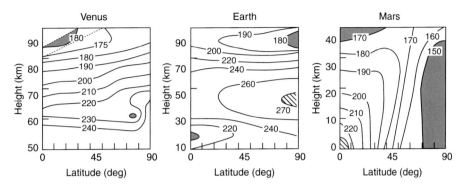

Figure 13.3 Meridional temperature cross-sections for the atmospheres of the terrestrial planets, from data by (a) *Pioneer Venus Orbiter* Infrared Radiometer, (b) *Nimbus 7* Stratospheric and Mesospheric Sounder, and (c) *Mariner 9* Infrared Interferometer Spectrometer.

characteristic circulation (Figure 13.4). Little is known about the deep atmosphere except that the winds are sluggish and we expect change to be slow. In the middle atmosphere, the dynamical behaviour is dominated by the Hadley circulation, the zonal super-rotation and the polar vortices. These all have their equivalents on Earth, but with major differences. The Hadley circulation reveals itself on Venus in movies of the observed migration of the ultraviolet markings away from the equator in the meridional direction towards both poles, at gentle speeds of a few metres per second. The general impression is of two gigantic circulation cells, one to each hemisphere, in which air travels from the equator to the edge of the polar vortex, polewards above the clouds and (by inference, since mass must be conserved) equatorwards below.

This type of circulation was proposed by George Hadley as long ago as 1735, to explain the trade winds in the Earth's atmosphere. Although Hadley himself would not have dreamt it, his is a logical circulation regime to expect for Venus where the Sun is always above the equator to within a couple of degrees, and the solid planet rotates slowly. The air warmed at low latitudes rises and moves towards the poles, where it cools by radiation to space and descends before returning equatorwards at lower altitudes. On Earth, which rotates relatively rapidly, the Hadley cell extends only to mid-latitudes in each hemisphere, with smaller cells taking over the transport towards the poles.

On Venus, the cell extends right to the polar vortex, feeding it with angular momentum from the equatorial super-rotating atmosphere. Some of the evidence for this comes from measurements of carbon monoxide, which is present in the deep atmosphere in concentrations of only about 30 parts per million. There is a large source of the monoxide in the upper atmosphere, as a result of the dissociation of CO_2 by solar radiation, so high concentrations of CO lower down, as first observed by *Galileo*, suggest extensive downwelling. More detailed CO measurements by *Venus Express* (Figure 13.5) show that the CO below the clouds peaks strongly around the outside edge of the polar vortex, marking the poleward edge of the Hadley cell.

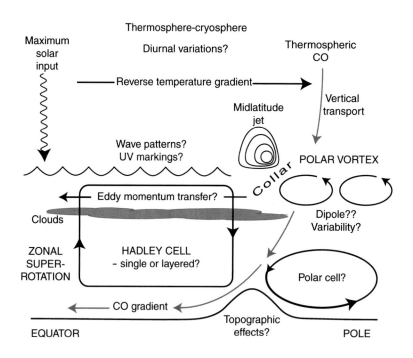

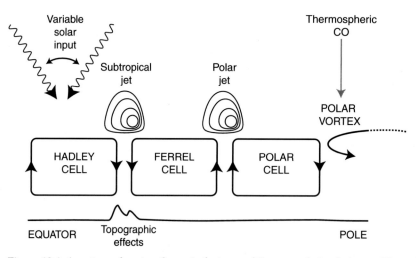

Figure 13.4 A cartoon showing the main features of the general circulation on Venus (above) and Earth (below).

Explaining the zonal super-rotation

On top of the simple overturning motions of the Hadley circulation are superimposed various complications, most conspicuously the global super-rotation at the cloud tops, consisting of high winds that move rapidly around the planet in a direction parallel to the

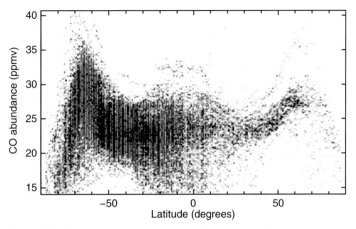

Figure 13.5 The carbon monoxide abundance below the clouds at a height of about 30 kilometres above the surface, as inferred from infrared spectra obtained by *Venus Express*.

equator. As we have seen, the cloud markings, which appear with high contrast through an ultraviolet filter, have their origin in the upper cloud layer, where the pressure is not much less than 1 atmosphere, and travel around the equator in four days, corresponding to speeds near 100 metres per second. Measurements of the winds below the clouds, and calculations (from temperature data) of the winds above the cloud tops, show that the zonal wind speed declines at higher and lower levels, reaching values near zero at the mesopause and near the surface respectively.

The deceleration that takes place above the clouds seems to be associated with the pressure gradient resulting from the temperature distribution at those levels, in particular the fact that the air temperature is nearly 20 degrees warmer at the pole than at the equator in the middle atmosphere. Dynamical models affirm that this type of gradient is sufficient to arrest the zonal winds completely by an altitude of around 90 kilometres. Below the clouds, the winds fall gradually in velocity as the atmosphere becomes denser and drag increases, and are close to zero at the surface.

All of the zonal winds are westward (in the same direction as the rotation of the planet), as we would expect if angular momentum were being delivered to the atmosphere by the solid body of the planet and transported upwards. An alternative mechanism is that the Sun exerts a torque on the atmosphere and drives the winds, supplying external angular momentum. This it certainly will do, since the density of the atmosphere is non-uniformly distributed with solar longitude (local time of day) because of thermal tides induced by solar heating. In fact, the semidiurnal component of the tide, on which the torque principally is exerted, has been observed to be unexpectedly large on Venus, relative to the diurnal component, which favours this mechanism. Whether the effect is large enough to accelerate the atmosphere to the speeds observed has been a subject of much debate. There is even conjecture that the slow retrograde rotation of the planet itself may have been established, over geological time, by the torque which the atmosphere exerts on the planet – the reverse of the earlier theory.

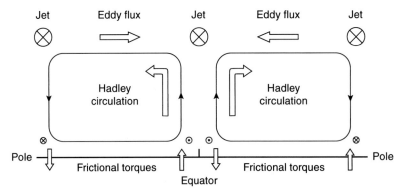

Figure 13.6 The basic mechanism by which angular momentum may be transferred from the surface to accelerate the winds in the middle atmosphere, producing the global super-rotation.

Current opinion favours a version of the mechanism in which momentum from the solid planet is transferred to the atmosphere by friction and then transported by waves (Figure 13.6). Exactly how these interact with the main flow is complex, but probably the mean meridional circulation and the solar tides play an important role. Computer models have been devised which, with some difficulty, are able to produce large zonal velocities in the Venusian atmosphere, although of course this is not the same as saying that we understand the forcing or dissipation mechanisms responsible for the transfer of momentum from the surface to the cloud tops. For the most part, these are inserted into the model artificially, a procedure called parameterisation. This is not as bad as it sounds, since we know such processes must be at work and the fact that we do not know enough to model their details explicitly is regrettable but not fatal.

Considerable insight can be gained by trying to simulate the observed behaviour of the atmosphere with computerised global or general circulation models. Usually abbreviated GCM, these are widely used for weather and climate forecasting and meteorological research on Earth. The familiar dire warnings about upcoming global warming, icecap melting and flooding come from the predictions of many and various GCMs, as does our daily weather forecast. They work by solving the equations that govern the motion of the atmosphere, starting with a set of initial conditions for the temperature profile, cloud properties and so on. The equations are represented on a grid of points in the horizontal and vertical, including numerical schemes that conserve key properties including energy and angular momentum, and incorporate approximations to represent processes such as turbulent eddies that cannot be explicitly resolved.

The first planet other than the Earth to be the subject of this type of study was Mars in the late sixties, during the very early history of terrestrial GCMs. The first attempt to do the same for the more difficult case of Venus was made around ten years later. The crucial difference between the two planets is that on the Earth the diurnal heating cycle is mainly at the ground, which tends to tie the forcing of the atmosphere to the rotation of the planet. On Venus, the energy from the Sun is mostly deposited in the clouds, well above the surface, and the two are more nearly decoupled, allowing the atmosphere to run away, in a sense, from the surface.

Developing this simple idea in detailed numerical models has turned out to be far from easy. In the first Venus GCM, the main result was that the simulation of strong prograde equatorial winds compatible with the observed super-rotation of Venus required the addition of a large ad hoc factor assumed to represent the pole-to-equator transport of angular momentum by eddies. This was not only somewhat arbitrary, but also led to problems with conservation of angular momentum. Later versions included a diurnal cycle, adding momentum transport by thermal tides to a better parameterisation of more realistic eddies. However, this achieved only moderate equatorial super-rotation, and the main conclusion was that tides acting primarily within and above the cloud layer on Venus can account for at most about half of the observed super-rotation.

The problem the early modellers were having was that they were calculating statically unstable radiative equilibrium temperature profiles. This means that the temperature gradient was, as it does on Earth and Mars, driving significant amounts of vertical convection. The resulting overturning of the lower atmosphere has the effect of supressing super-rotation near the tropopause. Once results from the *Pioneer Venus*, *Venera* and *Vega* probes had shown that the Venus atmosphere is statically stable above an altitude of 5 kilometres except for isolated near-neutral layers near 30 and 50 kilometres, a considerable step forward was possible.

The probes showed that the cloud intercepts most of the residual incoming solar flux, the part that is not reflected back to space. This heats the atmosphere near the top of the troposphere and produces a statically stable radiative equilibrium state in the lower and middle troposphere. The 'top-heavy' radiative heating profile that was measured, when introduced into the models, limited the depth of the convective region and helped to detach the upper-level flow from the surface. With this improvement, the GCMs could produce equatorial winds similar to the observed super-rotation, although there was still a problem in that they had to assume very low levels of stratospheric drag. Drag cannot really be ignored, especially in a model that relies on strong eddy activity to transport angular momentum, and when it was included in the simulations, the mean wind speeds fell by around half. An experiment in which the cloud was removed from the model showed that this did indeed decrease static stability, increase vertical convective mixing and almost completely eliminate the equatorial super-rotation.

These model experiments showed, above all, that it is crucial to get the calculation of the radiative heating and cooling profiles correct. However, computing radiative transfer in the atmosphere of Venus is quite challenging. The very long paths of strongly absorbing, very dense, nearly pure CO_2 gas have no equivalent in the laboratory, nor in the atmosphere of the Earth, so we have to rely on theoretical calculations of spectral line strength and shape. As in the discovery of the near-infrared windows, where the same theories had originally shown that the windows should not exist, large uncertainties can be introduced by the inadequacies of our knowledge of spectral line shapes and strengths. On top of errors in computing the gaseous absorption are those introduced by modelling the scattering by clouds of largely unknown structure (macrostructure as well as microstructure), composition and variability, at visible and near-infrared wavelengths.

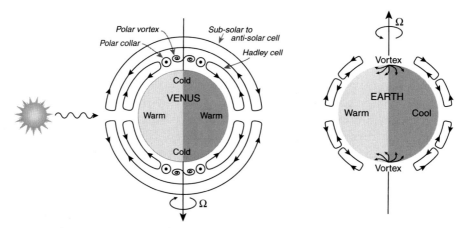

Figure 13.7 This simplified sketch of the main features of the zonal atmospheric circulation on Venus and Earth shows that a Hadley circulation and polar vortices are common to both. Earth has a secondary cell at mid-latitudes that seems to be absent on Venus, probably because of the slower rotation of the surface and lower atmosphere.

Even line-by-line codes,[5] tedious but generally accurate schemes that have been developed for the purpose of interpreting observations of the Earth's atmosphere, must be used carefully under the novel conditions of high temperature and pressure on Venus. The most accurate schemes are very expensive in terms of computer time and have to be deliberately degraded and simplified in the derivation of a fast, numerically tractable code for a GCM in which radiative transfer can be computed every few minutes if necessary, over a broad spectral range and for vertical profiles associated with each point of the horizontal grid.

The polar vortices, collar and 'dipole'

Vortex behaviour occurs in the polar region of any Earth-like planet because of the subsidence of cold, dense air at high latitudes, and the propagation and then concentration of zonal angular momentum in the meridional flow. On Venus, the small obliquity and the large super-rotation lead to an extreme version of this effect, characterised by a sharp transition in the circulation regimes in both hemispheres at a latitude of about 65 degrees (Figure 13.7). This is where the Hadley cell stops and we find the circumpolar collar, a belt of very cold air that surrounds the pole at a radial distance of about 2,500 kilometres. Measurements, and eventually models, showed that the collar has a structure like a single-maximum wave, locked to the Sun so that the coldest temperature stays at the same local time of day. There is no obvious reason for that, but the likely explanation is that the single-maximum component of the solar tide propagates more efficiently polewards than the double-maximum or 'wavenumber-2' component that dominates at the equator.

---

[5] They are called 'line by line' because the contribution of each spectral line is computed separately by numerical integration. For faster schemes, a statistical sum is used instead.

The vertical extent of the collar must be much less than its 5,000-kilometre diameter, and it may be only about 10 kilometres deep. The temperatures of the air in the collar are typically 30 degrees colder than at the same altitude outside, so the feature generates pressure differences that would cause it to dissipate rapidly were it not continually forced.

Cloud-tracked winds reveal that the collar marks the transition between a low-latitude regime, which conserves angular momentum, and the vortex, which tends more to solid body rotation. Inside the collar, the air at the centre of the vortex must descend rapidly to conserve mass. We should expect to find a relatively cloud-free region at the pole, analogous to the eye of a terrestrial hurricane but larger and more permanent, as a result of the downward flow suppressing cloud formation.

This is indeed the case, but interestingly the 'eye' of the Venus polar vortex is not even approximately circular, but elongated, with brightness maxima at either end of a quasi-linear feature connecting the two. In the low-resolution infrared maps obtained by *Pioneer Venus* over the north pole in 1979, this gave the eye a 'dumb-bell' appearance and led to the name *polar dipole* for the feature (Plate 8). From 2006 onwards, *Venus Express* made detailed observations of the corresponding phenomenon at the opposite (south) pole, and found that the dipolar structure is only one of many manifestations of a very complex phenomenon (Figure 13.8).

The northern dipole was observed in successive images to be rotating about the pole with a period whose dominant component was 2.7 Earth days, corresponding to about twice the angular velocity of the equatorial cloud markings. If angular momentum were being conserved by a parcel of air as it migrated from equator to pole the dipole might be expected to rotate five or six times faster than this. In fact, as we have seen, the ultraviolet markings follow a roughly constant wind speed from the equator to at least 60 degrees of

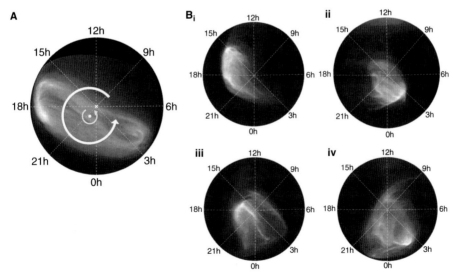

Figure 13.8 The many faces of the polar 'dipole', in infrared images from *Venus Express*. Although the name has stuck, it is clear that more complicated shapes can be seen now that we have sufficient resolution. For example, the image at bottom right is more like a tripole.

latitude, which corresponds to a fall in the rotation rate for mass motions around the pole. While some or all of its apparent rotation could simply be the phase speed of a wavelike disturbance, the dipole may also be the optimum configuration to transport angular momentum downwards at the pole. It is intuitively obvious that a rapidly spinning double vortex could do this more efficiently for a given mass of descending gas than a simple vortex could.

The south polar dipole is much better studied than the northern one, because of the long lifetime of the *Venus Express* orbiter and its more advanced instrumentation, which improved the spatial and time resolution of the data enormously over the earlier *Pioneer Venus* mission. Although there are no detailed observations of the north and south vortices made at the same time, everything we see in the two data sets separated by 30 years is consistent with the phenomena at both poles being very similar. The shape changes illustrated in Figure 13.8 are accompanied by rotation at a rate that varied between 2.2 and 2.5 rotations per (Earth) day for the south, compared with 2.7 for the north.

The centre of the structure, although not easy to define given the complex and changing shape, is often displaced from the rotation pole of the solid planet, by about 3 degrees on average, and precesses around it with a period of five to ten days. The precession period itself oscillates with a period of about three days. The reasons for this behaviour remain to be worked out, but the periods must reflect some fundamental modes of the atmosphere that dynamical models will eventually elucidate. In the meantime, the implication that the atmospheric circulation is not axisymmetric has important implications for understanding the super-rotation, since it offers a way to transport eddies from pole to equator and balance the angular momentum budget.

A 'strange attractor' on Venus

Their observation by the *Pioneer Venus* mission in the 1970s and *Venus Express* in the 2010s tends to confirm that the polar vortices on Venus are permanent and more intense than the equivalent phenomena on Earth or Mars. This fits with expectations from the evidence for the Hadley cell and for sustained zonal super-rotation. The curious double-eye in the cloud structure at the centre of the vortex also led to a theoretical analysis of the wave modes expected to develop at the poles under Venus-like conditions. Gratifyingly, this showed they are dominated by a wavelike instability with two maxima.

Together, the observations from the two Venus satellites revealed the details of the 'dipole' structure as a pair of giant, coupled vortices, rather than the simple analogue of terrestrial hurricanes that was originally pictured. Another piece of the puzzle may have been found when the resemblance of the S-shaped pattern formed by the compound vortex to the 'strange attractor' formed in the world's first computerised climate model,[6] by Edward Lorenz at MIT in the early 1960s, was noted (Figure 13.9). This has to be at least

---

[6] This is the famous 'butterfly' diagram that predicts two linked climate regimes, between which the Earth flips spontaneously. Although too simplified to be considered any kind of a forecast, it did inspire thinking about possible multiple stable states in the real climate of the Earth. It is an attractor because both regimes have a central value towards which the climate is drawn, and strange because the climate never has the same value twice.

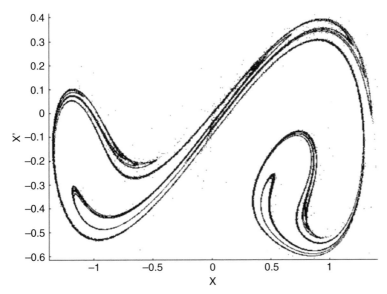

Figure 13.9 A 'Poincare section' through the Lorenz strange attractor. The resemblance to the most common shape of the polar 'dipole' on Venus is remarkable.

partially something of a coincidence, but does show that the exotic appearance of the polar dipole and its behaviour, which caused so much head-scratching in the 1980s when it was first observed, can be reproduced with basic equations representing atmospheric behaviour.

Waves and cloud features

The tracking of meteorological features – fronts, cyclones, waves and so forth – in images of cloud fields obtained from orbit is a well-established part of terrestrial research and forecasting. The Venusian equivalent was, for many decades, limited to the transient and quasi-permanent features seen in the ultraviolet images of the cloud-top region, which revealed structures identified with planetary wave activity.

The most pronounced of these is the large 'sideways-Y' feature, which cameras on Earth-based telescopes equipped with ultraviolet filters have been able to monitor for nearly 100 years. The nature of the Y-shaped marking became clearer from a mosaic of *Mariner 10* ultraviolet photographs taken during the flyby in February 1974. It is apparent that a large propagating wave having a wavelength equal to the equatorial circumference of the planet is present, and, by analogy to terrestrial phenomena, involves the superposition of two planetary-scale waves, one dominant near the equator and the other at the mid-equatorial latitudes.

Among the other important wave phenomena on Venus (Figure 13.10) are the circumequatorial belts. These are very narrow (less than about 50 kilometres in width) features, of variable length (of the order of thousands of kilometres) and transient appearance. As many as five have been seen at once, evenly spaced by about 500 kilometres and

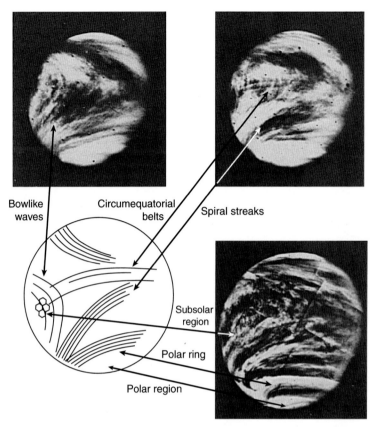

Figure 13.10 Photographs of wave features observed by *Mariner 10*, with an interpretive diagram by the science team for the mission.

always aligned parallel to the equator. Once they appear they propagate, always in a southerly direction, for several days at about 20 metres per second.

The most satisfactory explanation for the belts is that they are a form of gravity wave, that is, resonances caused by density variations propagating as waves under the influence of gravity as the restoring force. They are common in the Earth's atmosphere, and indeed they have been observed in satellite pictures of terrestrial clouds because temperature fluctuations, associated with the density waves, lead to condensation in the thermal troughs. Something similar may be happening on Venus. It is far from clear, however, what is exciting the Venusian waves. It could be turbulence in the strongly heated subsolar region, or perhaps some lower atmospheric wave propagating upwards. However, it is difficult to explain why the waves always seem to travel from the north to the south.

The bowlike waves were named for their shape (like that of a bow, as in archery). However, it turns out that they probably have something in common with the bow waves associated with the passage of the bow of a ship across water. On Venus these are probably related to the powerful 'boiling' of the atmosphere at the region directly below the Sun, near local noon. The rising of the heated air, visible as convection cells in the images, interferes with the smooth, high velocity flow of the upper atmosphere and generates 'ripples' in the clouds. However, this explanation is even more limited than its obvious oversimplification would imply, because the waves travel downstream behind the subsolar zone, whereas oceanographic bow waves remain fixed with respect to the disturbance producing them.

The pattern of cellular features in the subsolar disturbance suggests strong, localised convection. The two types of cells seen (light with dark centres, and vice versa) have been likened to the open (descending at the centre) and closed (ascending at the centre) cells in convective regions of the Earth's atmosphere. Scaling from their terrestrial counterparts suggests that the Venusian cells are about 15 kilometres deep, and the wind shear does not exceed 2 or 3 metres per second per kilometre. This in turn implies zonal (east–west) velocities of 50 metres per second or more at the bottom of the active layer, which seems reasonable.

The thermal tide (the diurnal increase and decrease of temperature caused by the rising and setting of the Sun) around the equatorial regions on Venus has two maxima and two minima, as often seen in the polar vortex. The two do not seem to be directly connected, because the two regions are separated by a narrow latitude band apparently free of planetary-scale waves, as well as by the predominantly wavenumber-1 collar. In the Earth's atmosphere, wavenumber-2 is only a small component superposed on the familiar early afternoon maximum to post-midnight minimum cycle, but it dominates on Venus. The dynamical theory of atmospheric tides, as developed for Earth, obligingly shows that when this is applied to Venus the observed state of affairs is to be expected, and can be explained as primarily a consequence of the long solar day on Venus.

Deep cloud motions and upper atmosphere dynamics

The best observed part of Venus's atmosphere is between about 50 and about 80 kilometres above the surface. This is the regime where the pressures and temperatures lie in about the same range as the Earth's atmosphere. Lower down, and higher up, data are harder to come by and it is less clear what is going on.

The deep atmosphere had been expected to be mainly quiescent, because the high densities lead to a long time constant for any response to changes in temperature. In particular, the calculated response time to the diurnal cycle of the Sun is about 100 times longer than the length of the solar day, even for the slow rotation of Venus. The winds near the surface are small but not zero, and are certainly large enough to produce topographic effects as they flow over the mountain ranges. The summits of the highest features on Venus are over 10 kilometres, where the measured wind is around 10 metres per second.

The dramatic-looking features in the lowest cloud layer, which are seen in near-infrared images such as that in Plate 7, are at quite high altitudes, nearly 50 kilometres, but well inside the troposphere, and representative of its meteorology. It is frustrating that we cannot see how deep they extend, and thus how they are produced. In Chapter 11 on volcanism, we saw that various lines of evidence suggested that the deep cloud features are in fact plumes of mineral dust from erupting volcanoes. If this turns out not to be the case, then some other explanation is required for the relatively compact nature of the features and their high contrast, with very opaque cloud surrounded by regions of nearly clear air.

The behaviour of the upper atmosphere, at heights more than 100 kilometres above the surface, is no less strange. There are no clouds there to track, and the air is too thin for following descent probes. This leaves spectroscopy, or drag experiments with satellites, as the only tools for probing the region. These techniques show that the thermosphere of Venus is cooler than Earth's, which is not surprising despite the closer Sun because of the greater abundance of carbon dioxide, which is very efficient at radiating heat to space.

Above about 150 kilometres, the temperature is approximately constant with height on the dayside at about 20 centigrade. The terrestrial thermosphere is the seat of rapid winds, as high as 1,000 metres per second, and this tends to redistribute energy originally absorbed from the Sun over the dark as well as the sunlit hemisphere. The day–night temperature contrast is about 200 degrees, whereas the mean temperature is nearly 750 centigrade. On Venus, however, the night-time temperature in the thermosphere is very low, around minus 200 centigrade. The transition from the dayside to nightside and back involves remarkably steep gradients (Figure 10.8), and modellers have great difficulty in reproducing both the minimum temperature and the short distance across the terminator at which it is attained.

Traces of oxygen, ozone and nitric oxide are produced by the action of the intense, short-wave solar radiation on the atmosphere at great heights. Production is highest at local noon, where the radiation is most intense, and results in some of the molecules being in an excited state. After a time ranging from microseconds to hours or even days, the excitation relaxes with the emission of a photon at one of several characteristic wavelengths. This is the phenomenon known as airglow, which is well studied on Earth.

On Venus, the observations from *Venus Express* show that the longer-lived excited species such as atomic oxygen collect at local midnight above the equator (Plate 20). This shows that, unlike the cloud tops far below, which spin around the equator and migrate towards the pole, the dominant circulation at these levels is from the subsolar to the anti-solar point, probably travelling over the poles as well as around the equator. This is the sort of circulation that is expected on a planet that rotates extremely slowly or not at all, and, before there were any relevant observations, it seemed possible it would apply at all levels to slowly rotating Venus. This is clearly not the case, however; below 100 kilometres or so the slow spin of Venus is still enough to constrain the circulation to the Earthlike Hadley overturning described earlier in this chapter. However, at great heights, where the densities are very low and solar forcing is large, the subsolar to anti-solar regime takes over. There must be a very interesting transition region at an altitude of between 80 and 100 kilometres, where the directions of the winds change and in some places reverse completely (Figure 13.7). Unfortunately, these heights fall in between the ranges covered by currently available wind-measurement techniques and have hardly been probed at all.

Chapter 14
The climate on Venus, past, present and future

Let us begin with a thought experiment. Had Venus and Earth been swapped at birth – that is, at the time when they had accumulated virtually all of their present mass but before their atmospheres were fully evolved – what would the inner Solar System look like today? In this thought experiment, Venus is now at Earth's distance from the Sun, and Earth 30 per cent closer than it was. Venus still rotates slowly and has any bulk compositional differences it acquired by forming at the closer position to the centre of the protosolar cloud, or as a result of any random processes that actually happened when planetesimals were combining to form the planets.

Currently, theories of the formation of the Solar System would have the gases in the protosolar cloud dissipate into space, and the atmospheres of Earth and Venus form later, mostly from gases that exhaled from their interiors. This would not have changed in their new positions, and in any case we believe at present that the gases that were supplied to the atmospheres of both planets were roughly the same. The motions of the planetesimals within the accretion zone jumbled the condensed and trapped volatiles that would later form the atmospheres of each planet. If Earth and Venus were truly identical in composition at the outset then presumably the result of swopping orbits over 4 billion years ago would be to produce much the same result as we have today.

However, other factors may be important. For instance, the small difference in size, some relatively subtle difference in core composition, or even (although most experts say not) the slow rotation of Venus, may all be responsible for the absence of an internally generated magnetic field. The missing magnetic shield against the solar wind could (although again this is being questioned) have been a key factor in the loss of water from early Venus. If it were, then our imaginary swop might produce a hot, arid Venus in Earth orbit, and a temperate, oceanic Earth in the orbit where Venus is in reality.

Some have argued that the planets formed fast enough to trap significant amounts of gas from the solar nebula, while others believe that a steady flux of icy, cometlike objects has modified their atmospheres over long periods, right up to recent times. In this case, our transplanted Earth might have had much more water than Venus from the outset and still have oceans despite being nearer to the Sun. Venus would be cooler than now, but still very hot due to having kept most of its carbon dioxide supply in the atmosphere.

None of these possibilities can be ruled out, since we cannot observe Venus at any time other than the present and the usual keys to the past, particularly the geological records, are mostly inaccessible for the time being. But we can consider the evidence we do have and try to construct hypotheses for how Venus's climate has evolved. Was it Earthlike in the past, as we believe Mars was? Will Venus naturally evolve to a more Earthlike state in the future (as Mars won't)? In the heyday of ground-based planetary astronomy in the first half of the twentieth century, it used to be fashionable to picture Venus as a kind of primitive 'pre-Earth' that was yet to evolve, and Mars as a played-out 'post-Earth' that had been fertile but then declined and died. Interestingly, the latest research with spacecraft is tending to bring that idea back. In this chapter we look at a modern paradigm.

Early Venus

In the beginning, huge quantities of carbon dioxide and water vapour were outgassed from the interior of the solid body as it cooled. The water was present as steam and made up most of the atmosphere, exerting about three times as much pressure at the surface as the carbon dioxide and nitrogen, making a total of nearly 400 atmospheres.

The high concentration of water vapour in the upper atmosphere, exposed to energetic solar radiation that can dissociate H_2O, leads to large amounts of hydrogen and oxygen as H and O atoms and ions on the fringe of space. The hydrogen boils off easily, its thermal energy alone being sufficient to escape Venus's gravity because of the small mass of the hydrogen atom. Most of the oxygen streams off into space as well, but not by thermal escape as it is 16 times heavier than hydrogen, which puts it under the threshold. Instead, it is carried away in the flow of hydrogen, and by the stream of energetic particles from the Sun in a combination of simple collisions and charged plasma interactions.

There is a problem during this epoch with the evolution and variability of the Sun. At present evidence is accumulating that most of the escaping gas is removed from Venus during certain intense episodes such as coronal mass ejections, known as CMEs, when the solar wind is much stronger for a short time.[1] Furthermore, in the distant past, when we would like to be able to calculate the rate of water loss from Venus in the days when it probably had a lot more, the behaviour of the Sun was probably quite different in ways that are hard to know. The best we can do is to make theoretical models of the Sun and test them against observations of stars of similar kind in various stages of their evolution. One of the conclusions of such research is that the Sun, like most young stars of its type, went through a 'T-Tauri' phase that was like a massive and more continuous CME.[2] This happened before the planets formed in their present state, and it removed most of the gas from the cloud of dust and debris surrounding the protosun.

---

[1] Strong enough to be fatal for any astronauts on their way to the Moon or Mars, if they are not well protected. The *Apollo* astronauts were fortunate that their flights took place when there were no large CME events on the Sun.

[2] The T-Tauri phase is named after the best-known example, and one of the first to be studied, of a star younger than the Sun that is in that state now.

Volcanic Venus

After perhaps a billion years of thermal escape and solar wind erosion the water on Venus is nearly all gone, and the rate at which it continues to escape is close to equilibrium with the supply from the surface. This supply comes from thousands of volcanoes, some giant cones, others just cracks and vents of various sizes, all belching gas and dust at a prodigious rate that is much higher than present-day Earth. Enormous amounts of lava also flow downhill to the lowest places on the surface, which would have been seabeds if Venus had been further from the Sun. Now they are filled with oceans of lava. Some of the flows were very copious and rapid, when the lava had a composition with a low viscosity at the prevailing surface temperature of around 1,000 degrees centigrade. Other flows also involved enormous amounts of very fluid lava, but more confined and over longer periods, so that they formed distinct sinuous river valleys that extended great distances from the valleys to the plains.

Gradually, the emissions decline, as the volatiles in the parts of the crust that are connected to the surface are depleted and the interior of Venus cools. Water, in particular, is baked out of the crust by the high temperatures. The continued heating by radioactive elements in the core also declines as its age approaches the half-life of the uranium and thorium components that are primarily responsible for releasing the heat.

Present Venus

After another 3 billion years we arrive at the time about 500 million years ago when the cooling of the core triggered a regime change. The rate of volcanic eruptions declines sharply and the remaining heat flux is mostly by mantle convection, a slow overturning of the molten rock below the thin, solid crust. Lifting, cracking and rifting of the crust occur on a wide scale, and some of the cracks act as vents for a smaller number of still-active volcanoes. Their total emission of heat and gas is now about the same as Earth, but the gases are mainly carbon dioxide and sulphur dioxide, with very little water because of the dried-out mantle.

The mass of the current atmosphere represents a balance between emissions from the crust by volcanism, the chemical recombination of atmospheric molecules with the surface, and the atmospheric sources and sinks at the boundary between the top of the atmosphere and space. Icy material arrives continuously as cometary and meteoritic debris and this is likely to be the main source of water on present-day Venus. It is balanced by the loss of hydrogen and oxygen to space by dissociation and ionisation, followed by thermal escape and particle erosion by the solar wind.

Now that the flux of carbon dioxide from volcanoes on Venus is about the same as it is on Earth,[3] the surface pressure will double in about a billion years if there are no losses. Alternatively, if the current environment is stable, then CO_2 is being removed at a rate

---

[3] The mean volcanic flux of carbonic gases into the Earth's atmosphere is estimated to be about 300 million tons each year (3×10^{11} kg yr^{-1}).

equivalent to losing the entire CO_2 component of the atmosphere in a billion years. The loss of carbon from the atmosphere to space occurs at negligible rates because its affinity for oxygen is much greater than that for hydrogen, so it is present mostly as CO_2 or CO and both are too massive to escape efficiently.

Chemical combination with the surface removes CO_2, especially reactions with calcium and other silicates, which should be abundant, to produce carbonates and other rock types. This rate of loss is very slow compared with the terrestrial rate, which mostly occurs in aqueous solutions that are not available on Venus, despite the dependence on temperature and pressure which favours Venus. Laboratory measurements of these rates for some of the commonest minerals have shown that they appear to be in equilibrium with CO_2 under the current surface conditions. This will tend to prevent the temperature and pressure from rising, or falling, despite the addition or loss of gas from volcanoes.

We pause at this point to remember that this is conjecture, a 'straw man' scenario based on what we know but incomplete and uncertain. The real situation must be much more complicated, and we do not even know that the surface pressure on Venus is not wandering up and down in response to long-term changes in volcanic output and possibly other factors. But let us press on.

Sulphur dioxide is a reactive gas and we expect it, too, to be mopped up by minerals on the surface, and at a much faster rate than carbon dioxide. It should be heading for a stable state that leaves far less of it in the atmosphere than is observed. The present proportion of sulphur dioxide in the Venus atmosphere is 100 times higher than the equilibrium value expected from the most likely models of the surface composition and laboratory-measured reaction rates, and this could not be maintained without a source. The fact that huge amounts are present means that sulphur dioxide is out of equilibrium because it is being pumped up by the relentless supply from volcanoes.

The combined effect of everything is to maintain the high surface pressure and dry conditions that are together responsible for the extreme climate on Venus, as represented by the mean temperature and pressure profiles shown in Figure 14.1.

Future Venus

In the distant future, more than a billion years from now, volcanism will have declined to almost nothing, as on present-day Mars. This is inevitable since the core will cool and its radiogenic heating will decline to a low level as the radioactive elements decay. What will the climate be like then?

The first major effect of turning off the volcanism would be the loss of most of the atmospheric sulphur compounds, just a few million years after the supply dwindles. The Venusian sulphuric acid clouds have a short lifetime that is determined by the timescale for sulphur dioxide to bind with the surface carbonates, and they will soon vanish.

The loss of the clouds will tend to lower the temperature of the surface as their contribution to the greenhouse warming of the surface is removed. Dwarfing this, however, would be the *warming* effect due to the decrease in the albedo of the planet. The amount of sunlight reflected would fall to about half of what it is now, equivalent to

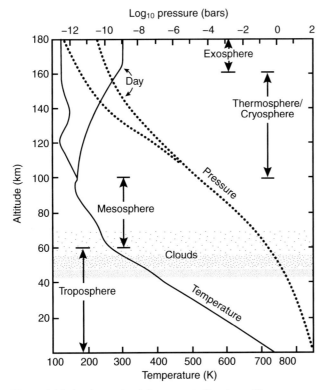

Figure 14.1 A schematic of the current climate on Venus, as represented primarily by the globally averaged profiles of temperature and pressure versus height, but also of course the atmospheric composition, including clouds.

turning up the solar input by a factor of two. Overall, the surface temperature will rise to a value that is even higher than it is now.

Eventually, however, this will be offset by the loss of atmospheric carbon dioxide to surface chemistry. Exactly how long this will take is particularly uncertain. If the calcite-wollastonite-CO_2 buffer dominates, it would hold the surface pressure and temperature to the equilibrium point for this reaction, which, as we have seen, is very close to the present values. However, the equilibrium is unstable, and a large perturbation such as the withdrawal of the volcanic supply could lead to slow but persistent CO_2 removal until after some long period of years the atmosphere consists almost entirely of nitrogen and the other chemically stable gases such as argon.

This is essentially the scenario that Arrhenius expected to find on Venus in his informed speculations about conditions there, which he published nearly a century ago. He overlooked the effect that massive volcanism, past or present, has had on the Venus climate by pumping up its mass, in the absence of the kind of efficient removal mechanisms involving liquid water which the Earth enjoyed. However, if, or rather when, the interior cools to the point where volcanism subsides, our paradigm suggests that the climate on Venus may evolve to a more Earthlike state, similar to that predicted by Arrhenius and others before the first surface temperature observations by ground-based and spaceborne microwave instruments, and the first direct measurements by landers.

Climate change models

We can put some numbers on this conceptual framework of the changing climate on Venus using simple models based on the relevant physics.[4] Figure 14.2 summarises the approach.

From a model of the surface composition (the Earth without oceans, essentially) we can calculate the rate of loss of carbon dioxide and sulphur dioxide by reaction with the crust. From assumptions about the interior evolution (again by analogy with Earth, but slightly smaller and with a dry crust) we can obtain the flux of gases into the atmosphere from volcanoes. Current knowledge of the solar wind and its history goes in to calculations of the rate of exospheric escape, and models of the Sun provide its intensity at the top of Venus's atmosphere.

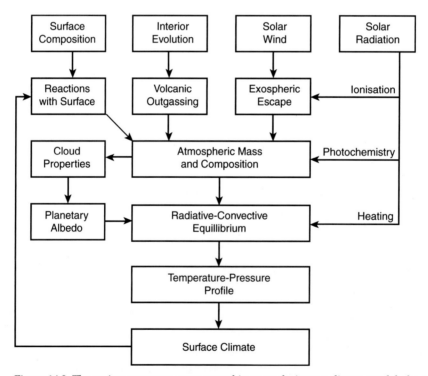

Figure 14.2 The various processes represented in an evolutionary climate model, the goal of which is to show how climate might change over time, past and future, based on measurements and constrained by basic laws of physics.

---

[4] The details belong elsewhere; see the References section at the end of the book.

The models have to be kept very basic so that artificial complexities are not introduced; just comprehensive enough for consistency with present knowledge and measurements. The aim is to summarise that knowledge in a framework of physical laws, in a way that permits coherent updating when new data or insights become available. Models are also useful for bringing out which factors are responsible for the largest uncertainties in the climate and the general circulation, and so defining the prime targets for study by future missions.

The integrity of the models is important; they have to be self-consistent as well as compatible with the relatively little that we know from observations. Calculated 'profit and loss' accounts for atmospheric gases are linked with what we believe about the composition of the primordial atmosphere on Venus and made to agree with what we measure there today with spectroscopic devices and mass spectrometers on entry probes. The condensable species, produced or modified by photochemical processes, form clouds in a roughly predictable way, and these have a calculable reflectivity or albedo that modifies the amount of solar energy deposited in the atmosphere.

The absorption and emission of radiation by the various gases and particles present are governed by formulae derived from quantum mechanics and molecular physics which, for the most part, are well known. The gases in the atmosphere are also subject to the laws of hydrostatics and thermodynamics, which again can be incorporated into the model, with the assumption that any atmosphere is approximately in the state known as radiative–convective equilibrium. Evaluating this predicts the temperature and pressure of the atmosphere as a function of height, and hence the surface climate as the lower boundary of this profile. Again, if the result of this calculation for present conditions does not agree with what we observe then the less certain assumptions that went into it need to be re-examined until it does.

Simplifications are possible at every stage, not only to save time and effort but also to avoid confusing complexity with precision. For instance, rather than carrying out very complicated and time-consuming radiative transfer calculations for a cloudy atmosphere in which basic parameters such as droplet size and shape are uncertain, meaning the result is uncertain as well, it might be just as accurate to adopt a plausible value for the cloud reflectivity as an input to the model. Some complexities are essential however, such as the feedbacks between different parameters. One important example of feedback is the effect that a change in temperature and pressure has on the rate at which reactions occur between atmospheric gases and the surface, as well as with each other. Another 'known unknown' is the time dependence of the emission rate of the various volcanic gases. This is sure to have been an episodic activity, but with an overall decline in the long term. Finally, exospheric escape depends on solar activity, which is likely to have been quite different when the Sun was younger, although these details are also quite uncertain.

Model forecasts and hindcasts

In Chapter 10, Figure 10.3, a simple model temperature profile for present-day Venus was shown to be a reasonable match to a measured profile. Most of the remaining discrepancy is due to heating by absorption of sunlight as it passes through the cloud layers, a process that the model disregards, and the fact that the model is meant to be a global and time average, whereas the details in measured profiles vary.

The model avoids the complexities of the radiative transfer calculation by adopting several reasonable approximations.[5] First, specifying the cloud albedo as an input to the model instead of trying to calculate it from particle sizes and so on. Secondly, by assuming an optically thin upper atmosphere (stratosphere) in radiative equilibrium, with the Sun above and the planet below. Finally, an optically thick lower atmosphere (troposphere) in convective equilibrium is represented by adopting the adiabatic lapse rate for the vertical temperature gradient, which is a function only of the molecular weight for a dry atmosphere such as that of current Venus.[6]

Since the model approximates the present climate to a useful degree, we can use it to consider how things might change in a future scenario, for example one without volcanism. The first consequence of such a change is expected to be the loss of the sulphuric acid cloud. We remove the cloud from the model by setting the albedo equal to that which we expect for a planet with a thick atmosphere without cloud. The atmospheric density is still high so the surface contribution to the planetary albedo would still be small, and molecular scattering by the atmospheric gases would dominate. The albedo calculated for this situation would still be higher than on Earth, since here half of the sunlight reaches the surface, which is a relatively good absorber, but less than on present Venus with its highly reflective sulphuric acid clouds. Among many uncertainties that we ignore is the possibility that water or some other type of cloud might form; these would probably be thin if they mattered at all in an atmosphere that would probably still be very depleted in water and other volatiles.

The model now predicts a surface temperature of over 600 centigrade; about 900 Kelvin (Figure 14.3). The figure also shows two model profiles for the even more distant future, where we hypothesise that most of the carbon dioxide has been removed as well. In one (labelled 'Bullock', because it was originally produced to compare to the models of Mark Bullock, see References section) the surface pressure is between 2 and 3 atmospheres, mostly nitrogen, and the temperature comes out close to 70 centigrade. We might expect nitrogen to be lost by weathering or solar wind erosion as well, but extremely slowly. If some of it goes, reproducing the scenario pictured by Arrhenius, the reduced pressure will lower the temperature further, with the intriguing possibility that conditions suitable for humans will prevail at last. Of course, we are talking of probably billions of years in the future.

---

[5] Calculations of the transfer of solar radiation through the atmosphere at all wavelengths, including the effect of absorption by all the gases and scattering in the clouds, require a complicated computer program that takes hours to run on a large machine. The corresponding scheme for infrared emission takes just as long. Those wishing to investigate the gory details, see *Radiation and Climate* by I.M. Vardavas and F.W. Taylor (Oxford University Press, 2011).

[6] On Earth, or early Venus, an extra term comes in to allow for the latent heat of water vapour, but the profile is still simple to calculate.

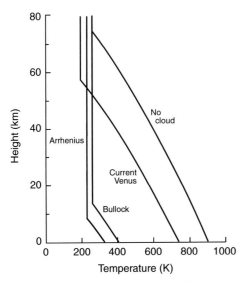

Figure 14.3 Model temperature profiles for present and possible future conditions on Venus. The three forward-looking scenarios have, progressively, the clouds removed; most of the carbon dioxide removed (Bullock); and just 1 bar of nitrogen (Arrhenius).

Summing up

Figure 14.4 sums up graphically the results of all this modelling of Venus's climate, with its four epochs. The key variable is the surface pressure and how it behaves as a function of time. In the early 'ocean' phase, if the amount of water was equal to what Earth has now but present only as steam, the surface pressure could have been in the region of 300 atmospheres. Once most of this has escaped, after perhaps a billion years, Venus enters the 'volcanic' era, where gases are emitted at a rate of up to 100 times the present-day terrestrial rate and the surface pressure declines but still remains higher than now, perhaps at 200 atmospheres. When the planet is about 4 billion years old, the interior enters the convective phase and volcanism drops to something resembling current Earth rates. The surface pressure falls as the activity declines, passing through the current value of nearly 100 atmospheres, as sulphur dioxide and carbon dioxide are lost to the crust. Eventually Venus cools to a point where its internal activity resembles current Mars, and very stable gases like nitrogen and argon dominate the atmosphere and produce surface temperatures typical of the tropics on present-day Earth.

Climate change on Venus over these long periods is inevitably paralleled by similar changes on Earth and Mars. Venus most likely did have an ocean's worth of water, but probably not in an Earthlike state, if by that we mean present-day Earth, and possibly never had any liquid on the surface, only steam. Of course, the Earth has evolved too; it is very likely that early Earth was hot, with a much higher surface pressure due to a lot more carbon dioxide than now. Our oceans would have been steam too, and water would have been lost to space more rapidly than now. The main differences from early Venus were

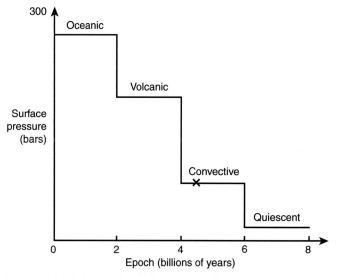

Figure 14.4 Model-predicted changes in surface pressure over time, showing the four epochs discussed in the text, with the present day marked by a cross. In reality the pressure is not expected to stay constant during each phase, nor to make steplike transitions between them, as shown; the simplified representation is deliberate to emphasise that the uncertainties remain very large.

probably that Earth's upper atmosphere was colder, tending to trap the hot water vapour below, and there may have been less erosion of the atmosphere by the solar wind because of the Earth's magnetic field.

Eventually, the Earth's 60 atmospheres (according to one estimate) of carbon dioxide was converted to carbonates and deposited as chalk in the features we see now. The atmosphere cooled and the water condensed into the oceans. Venus could eventually follow the same path, but too late to save most of its water. It could still come to resemble Earth much more closely than it does now, or did in the past, in the distant future when volcanism ceases.

However, there are so many ifs and buts about all of this that although the speculation may be fascinating it is not really very satisfying. When we know more about the nature of volcanism on Venus, and the chemical composition of the surface, things will be better, and with new missions waiting in the wings, hopefully this will not be too far off.

Another line of attack would be to find Earthlike and Venuslike planets orbiting other stars. Rapid progress is being made towards detecting such exosolar systems, and ultimately towards analysing their atmospheres and climates. What we will find will not be just analogues to Earth and Venus in their current states, but also younger and older versions in earlier or later stages of their evolution. Much of our present understanding of the Sun comes from this kind of approach, but since it is a lot harder to study far-off planets than their parent stars, we will have to wait a little longer for good telescopic surveys of conditions on Earthlike exoplanets, and longer still for interstellar space probes to provide the definitive answers. But we know enough already to be sure that such things will come eventually.

Chapter 15
Could there be life on Venus?

Is there life on Venus? Probably not.

We could leave it there, except for the fact that there are several good scientists who do not dismiss the possibility, so there is definitely something to discuss. Also, we have to admit that we do not really understand the conditions under which life can evolve, or at least survive, and accept that there is life on Earth in niches (inside nuclear reactors, for example, or deep down in cold, dark frozen lakes) where no one expected to find it until it was discovered recently. Certainly, if we want to know whether we are alone in the Universe, and most of us do, we should leave no stone unturned.

Possible habitats on Venus

There are two very different environments where we might look for life on Venus: in the clouds and on the surface. Most of the research to date has focused on the former, for the obvious reason that the temperatures and pressures near the cloud tops are close to those at the Earth's surface where we know that conditions suit life forms of many kinds. There is also a supply of liquid water, widely accepted as a prerequisite for life of any kind. There is also plenty of energy, as solar radiation beats down on the cloud tops with only moderate amounts of attenuation during its passage through the upper atmosphere. The vigorous dynamical activity, providing mixing and a strong diurnal effect, is all to the good. Finally, we already have evidence for chemical activity, for instance that which converts sulphur dioxide to sulphate, and this is no doubt only the tip of the iceberg as far as chemistry is concerned, and in a basically CO_2 atmosphere some of it will be pre-organic (at least), providing further sources of energy and the compounds that might be the building blocks of life.

Put like this, the prospects for life in the Venusian clouds begin to seem positively rosy. Of course, there is a downside. First, the water is in a solution of strong sulphuric acid, not a complete bar to life but certainly not ideal. Secondly, the solar radiation has a strong ultraviolet component that would be deadly to most known forms of life (and indeed in everyday life on Earth, materials and surfaces are often sterilised using ultraviolet lamps that are far less deadly than the flux at the cloud tops on Venus). And finally, there is no solid surface to provide a stable platform for the development of life. The amount of mixing that seems to go on in Venus's clouds would probably subject any bacterium riding in the droplets to a disconcerting amount of change in temperature and other variables. The cloud layers themselves are long-lived, on the grand scale, but if the lifetime of an individual droplet is short, before it evaporates and another forms somewhere else, it is hard to imagine the bugs hanging on.

We would always tend to look to the surface, or close to it either above or below, for the stable setting that life presumably needs. On Venus, of course, we run into the problem of the high pressure, and especially the searing temperature. Suppose we bypass the question of how life may have *originated* in such a setting, and consider the much easier question of what would happen to bacteria from Earth if they were introduced on Venus. This is a question that has been addressed in detail by experts, as part of the international quarantine regulations that space agencies voluntarily observe when planning to send man-made objects to (presumably) pristine environments elsewhere in the Solar System. It is a massive subject for Mars; even Jupiter was studied in this way before the *Galileo* probe was dispatched to plunge into the giant planet's atmosphere in 2003.

In 2006 the Task Group for Planetary Protection Requirements for Venus Missions set up by the US Academy of Sciences concluded that 'the prospects for the survival of organisms deposited by planetary probes on the surface of Venus are non-existent', citing 'the high temperature, the absence of water, and the toxic chemical environment'. They also decided that the strong oxidising and dehydrating effect of the sulphuric acid, and the ultraviolet flux, would make short work of any complex organics that were introduced at cloud-top level.

Since then, *Venus Express* has detected ozone in the upper atmosphere of Venus, raising the possibility of a protective screen against the most damaging ultraviolet rays, such as we enjoy on Earth. However, the amounts on Venus are tiny, and mostly concentrated near local midnight where the descending branch of the circulation in the upper atmosphere brings down atomic oxygen produced by the dissociation of CO_2. The maximum amount of ozone in a column is about one hundredth of a Dobson unit, around 1 million times less than on Earth, and not enough to provide any significant protection from ultraviolet for any life that might exist on Venus.

Evolution and adaptation

Another way of looking at the question of life is to start with the warm, wet, more Earthlike Venus that many of us think might have existed billions of years ago. It is by no means certain that there ever was such a climate on Venus, but if there was then the likelihood that life evolved then is much higher than if Venus has always been hot and/or always dry.[1] The challenge then becomes to understand what

---

[1] As discussed in earlier chapters, the more likely scenario is that early Venus was hot and wet, rather than warm and wet, so that although water was abundant, it formed a high-pressure steam atmosphere, which never condensed into a liquid ocean before most of the water escaped from the planet.

happened as Venus's surface environment slowly evolved to its present state. The most likely thing is that they simply died out, whatever 'they' were; we cannot yet say how far along the evolutionary path Venus life went, nor how much that path resembled Earth's.

However, we have come to understand that life is very persistent, and given benign conditions to get started and plenty of time to adapt (again, we cannot say what that timescale was), it may have found a niche somewhere on modern Venus. Acid-loving bacteria certainly exist, and some sort of protective coating against damage by ultraviolet radiation might have led to survival in the clouds. Possibly there are even ways to develop an affinity for temperatures above the melting point of lead, although biologists cannot see at present how that would work.

Terrestrial vs. Martian vs. Venusian life

Many of the questions we would like to ask about Venus life are still unanswered even for Earth, such as the probability for life to start up under given conditions, how long it takes, and how external forces drive evolution. We now know that life started very early in Earth's history, and that it is remarkably adaptable and durable. We also know, with more certainty than for Venus, that Mars had an Earthlike climate with liquid water on the surface in the past. The search for biology there is decades ahead of anything planned for Venus, quite rightly because the investigations are more feasible and more likely to deliver some sort of a positive conclusion.

However, we are also finding that Mars, and Earth itself, had very acid environments in these early epochs. Probably, the active volcanism that pumped up the atmosphere and hence the temperature on Mars, and supplied the water, was also responsible for emitting masses of sulphur compounds in much the way volcanoes do on modern Venus. Any Martian bugs will also have had to adapt to the strong ultraviolet flux at the surface of present-day Mars if they survived; another parallel with Venus. The analogy cannot be pushed too far, because the Martian subsurface is potentially an attractive habitat to which early life could have retreated, but the same cannot be said for Venus.

Perhaps the most important point when considering the trio of Earthlike planets is that we will know with reasonable certainty whether life arose on Mars, long before we have comparable results from Venus. If the answer for Mars is yes, then we will be a large step closer to believing that life arises easily when the circumstances allow, and we will know more about how it evolves to cope with more adverse surroundings. There may be hope for Venus then. If Mars is, and always was, barren, then we would need a very good reason, as yet unidentified, for taking a similar amount of trouble over the biological exploration of Venus in the foreseeable future.

Possible evidence for life

It was famously pointed out by Carl Sagan, many years ago, that it would be difficult for an observer from another planet to positively identify the existence of life on Earth from a

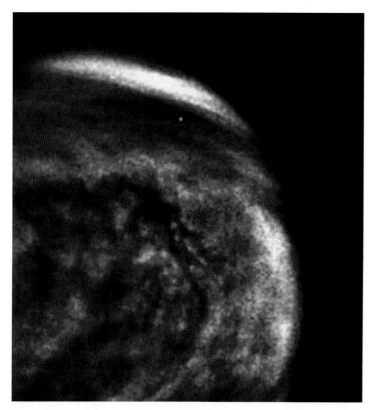

Figure 15.1 Some think the dark markings in the clouds could be colonies of ultraviolet-absorbing bacteria.

distance. Various studies (e.g. one undertaken after the flyby of Earth of the *Galileo* spacecraft, bound for Jupiter, in 1990) have concluded that a well-instrumented spacecraft would obtain only indirect clues – the oxygen in the atmosphere, for example – unless it gets very close or lands. Even after landing, intrepid alien astronauts would have to do some work, especially to discover advanced life, unless they were lucky enough to land in the middle of Chicago or similar rather than, say, the Gobi desert.

So what has been seen on Venus that might shed light on this difficult but important question? One thing might be the mysterious ultraviolet markings (Figure 15.1). We have tried to explain them in terms of inorganic chemistry, perhaps leading to the formation of various exotic forms of sulphur that absorb ultraviolet to match the spectrum of Venus and that might vary in the manner observed. However, this has not been fully explained, even theoretically, and there are as yet few supporting observations. This leaves scope for those who want to suggest that the markings are in fact carpets of bacteria, living in the cloud droplets and absorbing the energy from solar ultraviolet radiation for some advanced form of photosynthesis. A plausible metabolism has been suggested whereby sulphur dioxide and carbon monoxide are converted to hydrogen sulphide and carbonyl sulphide. It was ingenuously pointed out that, if bacteria were in fact doing this it would explain the otherwise curious balance amongst those gaseous species in Venus's atmosphere. It remains more likely that we just don't understand the inorganic chemistry well enough, but the biological alternative is intriguing.

Another angle on this is the suggestion that the ultraviolet absorber is not the organisms per se, but rather the material that they use to protect themselves from the sterilising effect of the ultraviolet flux from the Sun. A candidate for this is cyclo-octa sulphur, a very stable material with eight sulphur atoms forming a ring. This absorbs ultraviolet photons without dissociating, and could conceivably be manufactured by the organisms from the various sulphurous gases that certainly are present inside the clouds. A thin coating a few molecules thick would be enough to protect them, and at the same time produce the cloud markings whose origin has been an ongoing mystery for nearly 100 years.

To complete this section it is necessary to mention, and hard to resist including a picture of (Figure 15.2), the (very) tentative identification of (vaguely) possible life forms on the surface in some of the old *Venera 13* lander pictures that came to light recently. Blurry shapes, one of them (slightly) resembling a scorpion, are seen, and said to 'emerge, fluctuate and disappear' between successive photographs. The scientist who came up with this, in 2011, was Leonid Ksanfomaliti of the Space Research Institute in Moscow. He was one of the *Venera* scientists at the time of the landing (1982).[2] He is quoted as saying 'What if we forget about the current theories about the nonexistence of life on Venus, let's boldly suggest that the objects' morphological features would allow us to say that they are living.'

Astrobiological experiments: next steps

Now we ask what new measurements might be made to advance our future understanding of the prospects for life on Venus in the relatively short term. Many of the items on this list are not that different from the non-biologist's list, for example that which we would produce for future missions aimed at understanding the divergence in climatic conditions on Venus and Earth. Answers to all of them are probably a prerequisite for mounting any kind of direct biological investigation, such is the complexity of that kind of experiment, and the current scepticism in most quarters about whether there is anything there to find.

- How and when was the surface water lost? How are gases escaping from the upper atmosphere and how does this change in response to changes in the Sun?

---

[2] The author met Ksanfomaliti at scientific conferences in the 1980s and can certainly testify that he did excellent work then. Russians are not noted for tongue-in-cheek humour, but I suspect he is having a joke with the press in his old age. His hypothesis appeared in the journal *Solar System Research* (vol. 46, pp. 44–57, 2012, in Russian) and was released by the Russian State News Agency.

Figure 15.2 Top, the 'scorpion' seen on the surface of Venus by magnifying part of one of the surface panoramas (below) photographed from *Venera 13* in 1982.

Life as we know it needs water. If Venus had an early ocean of liquid water, the chances of life developing long ago and hanging on today in some niche underground, or in the clouds, is much higher. This is the sort of scenario currently considered appealing for Mars, and the question will likely be resolved there long before it is on Venus, because the definitive evidence for past oceans will lie in the sediments that were deposited billions of years ago. Now, these must lie well below the present surface, underlayers of volcanic lava, requiring extensive (and expensive) exploration techniques of the kind that are just about possible on Mars, using the next generation or two of mobile robots with deep drills, and eventually human explorers. On Venus, this is obviously not so simple.

• What is the history of the climate? Was it once wet and Earthlike? Are the clouds a stable niche for life?

Despite the fact that current thinking stresses the likely importance of 'extremophiles' – that is, organisms that thrive in difficult environments, including high temperatures – for the evolution of life on Earth, this almost certainly does not extend to the molten-metal environment on the surface of Venus. If there is, or was, life on Venus it would have needed a cooler environment than that found there today. So we need to know if it was cooler in the past, as well as wetter. If not, life would probably have to arise in the clouds, not just cling on there, a much less probable scenario. The clouds are very dynamic, with high winds and turbulence; not the sort of platform that would provide a stable niche for the development of life, even if all of the ingredients were there. There may have been a time when there were no substantial clouds at all, although that seems unlikely, particularly since volcanic emissions of water and sulphur dioxide are expected to decline, not increase, over time. On the positive side, the recent evidence that seems to confirm the existence of inter-cloud lightning strikes offers a source of energy that may always have been available for synthesising complex molecules, a process that is often cited as having been key to the origin of life on Earth.

• Are there clues to a past biosphere in things like noble gas abundances and isotopic ratios that tell us about the origin and evolution of the atmosphere?

We have seen that isotopic ratios in hydrogen (the D/H ratio) inform us about the history of water, and that studies of the small amounts of the stable 'noble' gases in the atmosphere constrain the history of outgassing and geological formation. On Earth, we have also been able to work out some of the biological history of the planet from things like the isotopic ratios in carbon and oxygen. Eventually, we can hope that similar evidence may be uncovered on Venus, although at the moment it is not entirely clear what to look for. For now, it is mainly a question of establishing the history of outgassing from the interior and trying to infer whether there was ever a habitable phase at the surface, while going on to make more and more detailed and precise measurements of the tiny traces of rare isotopes, especially in the atmosphere but also in surface minerals when we have samples to hand, and hoping to spot something unexpected.

• What minerals are on the surface? How old are the surface rocks in different regions?

We need to know a lot more about what the surface of Venus is made of in order to understand the stability of the present atmosphere. Then we would like to look in boreholes and cliff faces, for example, for layers that spell out the history of the surface and its chemical exchanges with the gases in contact with it. This will also reveal how benign an environment the surface was for life in the past. We already know that most of the surface of Venus has been covered in deep layers of molten lava relatively recently; not an ideal way to sustain life hanging on in an early ocean bed, for instance. However, relatively small regions of higher ground have apparently escaped this fate. We would like to know when those surfaces formed, whether liquid water was involved, and what record of the past they have stored by way of ancient deposits. Fossils would be great.

• What unknown trace chemicals exist in the clouds? What is the 'unknown ultraviolet absorber'?

Over much of its depth, the cloud material on Venus is mostly about 80 per cent sulphuric acid and 20 per cent water, bound together. The vertically thin, but optically very dense, layer at the base of the acid clouds has something else in it, most likely volcanic ash. The thin haze in the hot region between the main cloud base and the surface could be specks of solid sulphur. None of these is very encouraging as a life-supporting medium, except the water of course, but that is bound with H_2SO_4 making a very concentrated and corrosive solution. Extremophiles do exist that like a very acid environment, so that may not be fatal, although they would also need some sort of a shield from the Sun, since the solar ultraviolet radiation at the cloud tops, although attenuated by its passage through the thermosphere, should still be strong enough to guarantee a sterile environment. The dark material seen in the clouds may be evidence for such a shield, or even for exotic, robust microbes that use the energy of the photons to thrive. It is also the case, however, that a life form will need some kind of nutrients to live on that are more complex than the simple compounds of carbon, hydrogen, sulphur and oxygen that we already know about (CO_2, CO, H_2O and so on). Direct sampling with sophisticated equipment to search for complex organics and compounds of phosphorus, chlorine and iron is called for. If the results are encouraging, this could be followed by microscopic and other tests for the presence of actual microorganisms.

• How important is disequilibrium chemistry, for instance where the atmosphere meets the surface? What is the gas composition of the near-surface atmosphere?

Life is the ultimate disequilibrium process, so our search for life in the universe had better include understanding simpler examples first. One we have already discussed, that is potentially important on Venus, is whether the minerals in the surface are steadily gobbling up the CO_2 and SO_2 that is being emitted in copious quantities by lots of active volcanoes. The part of the surface pressure that is due to CO_2, that is, most of it, has the right value for this to be the case. SO_2, much less abundant, is nevertheless present in quantities much too large to be in equilibrium with a reasonable model of the surface composition. That does not mean that it would not tend to equilibrium if the source was turned off.

Once we know what is going on with these common and relatively well-studied species, we can ask questions about the equilibrium of lots of other atmospheric gases: chlorine and bromine compounds, for instance, or the biologically important compounds of phosphorus. The list is endless. And the surface is not the only place to look for action: there may well be significant levels of non-equilibrium chemistry in the clouds, involving a wide range of sulphur compounds for instance. If there are microbes there, they will contribute.

The prospects for finding life on Venus

It goes without saying that many of these objectives are very long term, some likely to require very advanced technology that enables sample return from Venus and manned laboratories in its atmosphere. In the shorter term, an automated *in situ* search for unusual and complex compounds and simple living organisms in the clouds is perhaps the fastest way to make progress on the exobiology of Venus. The problem remains that, for most of the planetary sciences community, especially those in a position to decide priorities, the prospects of success are so slender that they are unlikely to drive anyone's space programme for many years to come.

Part III
Plans and visions for the future

The last American mission to Venus was in 1991 (*Magellan*), and the last from the Soviets in 1985 (*Vega*). Europe has its first spacecraft to the twin planet still operating in Venus orbit at the time of writing, having arrived there in 2006 (*Venus Express*), but so far the Europeans show little political will to follow up on this success. Japan still hopes to get something from its *Akatsuki* probe when it meets Venus again six years after its failed attempt to enter orbit in 2010. What happens next?

In seeking to answer that question, armed with a mass of unsolved scientific questions from Part II, we must enter the murky and uncertain world of politics, budgets and planning, or 'programmatics' as it is euphemistically called in the space agencies. We are helped by the long-term planning process operated by the scientists in their institutes and universities, by the learned societies and national academies, and by the agencies themselves. But these produce wish lists (usually called 'roadmaps' these days) and not real predictions. Even the most optimistic planetary scientist knows that most of what they plan will not be carried out. So what will?

The existence of exciting and achievable scientific goals for Venus is a necessary, but not sufficient, condition for getting new missions approved and funded. Obviously, at a cost of at least several hundred millions of dollars or their equivalent, and sometimes ten times that, any new spacecraft is going to have to survive a competition for resources from other areas of science, and beyond. Even within space agencies, there is deep competition between disciplines (the new James Webb orbiting astronomical telescope versus sample return from Mars, to pick out one current example[1]), and even between planets, for the available funds.

---

[1] The successor to the wonderful and highly successful Hubble Space Telescope, the James Webb telescope is due to orbit the Earth and explore deep space in about 2018. Its estimated cost up to launch recently rose, to great publicity, most of it negative, from half a billion to over eight billion dollars. For comparison, Hubble also started at about 500 million and at launch had reached about five times as much. The largest planetary mission, the *Cassini Saturn Orbiter*, was estimated to cost just under one and a half billion and *Venus Express* about 250 million. These and similar figures always have to be treated with caution as there are many different ways to calculate them, and it is far from easy to decide what to include.

Venus is not very well placed in this competition. Driven by public interest in Mars, and the search for signs of life, NASA has ring-fenced a big part of its planetary budget for a stream of launches to the red planet, at least one and sometimes two each time the launch window opens, which is approximately every two years. While this pace is not likely to be maintained from now on, it is still the case that NASA's eyes are much more on Mars than they are on Venus. It has also become increasingly feasible to send large spacecraft to the outer Solar System, to investigate the gas giants and their families of moons, again with an exobiological carrot mainly in the form of Jupiter's moon Europa, with its warm water ocean beneath an outer crust of hard, cold ice. This has caught the European agency's imagination, with the Jupiter Icy Moons Explorer taking the lion's share of its planetary programmatic funding for many years to come.

Compared with these, Venus has been seen to have been largely 'done'; with the biggest scientific questions at least partially answered, and prospects for deploying the technologies we really need on Venus looking remote. The next steps, following the Martian example, would be to land robot rovers and to think about manned missions. However, the conditions the earlier missions found there would seem to preclude these steps for a long time to come. Furthermore, the chances of finding life are slim. Unlike Mars and Europa, no one would put their money on Venus missions contributing much to advancing the exciting science of exobiology. In the competition for resources to go back to Venus, our twin nearly always loses these days.

So what is going to happen? In the following chapters we look at the answer in three ways. First, overall strategy – how important are the remaining key questions about Venus seen to be, compared with everything else with a claim on the space agencies' budgets? Next, we look at active plans, proposals and detailed studies for future missions. These are projects that could actually happen, although they won't all get beyond the drawing board. Finally, future prospects for the very long term, beyond any realistic planning horizon. Dreams, if you like. You don't get anywhere without dreams.

Chapter 16
Solar system exploration: what next for Venus?

Choosing the way forward

When *Venus Express* arrived in April 2006 it became the 25th mission to target Venus successfully. With Japan's *Akatsuki*, four spacefaring nations are now engaged in exploring Venus, and the data garnered have painted a vivid and comprehensive picture of what hitherto had been a mysterious, cloud-shrouded world. No seas, swamps or rainforests and no dinosaurs or Treens, but a hot volcanic wasteland with permanent hurricanes and searing acid clouds.

There were, and still are, plenty of puzzles to solve concerning the nature of the surface and interior, and the behaviour of the thick atmosphere. More missions must follow. But what, how and when? The scientists, engineers, managers and politicians who will answer these questions and write the next chapter in Venus exploration cannot consider Venus as a solitary objective. The agencies face huge internal and external competition for resources, and must target the highest priorities if they are to satisfy their own scientific communities, not to mention their government paymasters. Often this means leaving lower priority destinations unprobed for long periods, as has already happened with Venus during the years from 1994, when *Magellan* shut down, to 2006, when *Venus Express* commenced operations. So what happens next at Venus depends not only on goals and priorities for that planet, but on where the excitement lies elsewhere in the Solar System and beyond, and where Venus fits in with the rest of the international planetary programme.

Of course, such considerations are not just about competition. Just as we learn about our Earth by studying its companion and near-twin planet Venus, we learn more about both by considering them in the context of the entire Solar System. We know that the planets all formed together, out of the same cloud of material, and we know that they evolved, and currently behave, in accordance with the same laws of physics. Any programme of planetary exploration must consider what to do next in terms of the priorities for all of the planets, moons, and minor bodies, as well as the Sun itself and the Universe beyond.

Venus is naturally most often compared with its three inner solar system siblings, the other rocky planets Mercury, Earth and Mars. The gas giants are less obviously analogues, although they do have solid cores made up of the heavier elements below their deep atmospheres. In the case of Uranus and Neptune, these cores are probably not too different in size from all of the Earth, or Venus, while Jupiter and Saturn have cores that are perhaps ten times more massive. Deep under thousands of kilometres of compressed hydrogen and helium, the parts of the gas giants that may be similar in elemental composition to the rocky planets are not only inaccessible to our probes but also in exotic states of matter because of the enormous temperatures and pressures they experience.

We can, however, explore the outer shell of each outer planet atmosphere, down to pressures approaching those at the surface of Venus, with orbiters and probes. All of them, most familiarly in the case of the nearest and best-studied giant, Jupiter, have some atmospheric properties that we recognise on Earth, such as cloud layers, lightning, and storm systems. But Jupiter and Earth are not immediately comparable, in the way Venus, Earth and Mars are, partly because the atmospheric composition of Jupiter is almost completely different from the terrestrial group. Also, the outer planet atmospheres are extremely deep, with no solid surface in the sense familiar to Earth dwellers. Finally, the mainly fluid nature of the outer planets gives rise to internal sources of heat that are similar or larger than the total input from the Sun, so the atmospheres are heated from below to a much greater degree and their meteorology has a very different character.

At the other extreme, in terms of mass and atmospheric density, lie the smaller bodies of the Solar System, the moons, asteroids and comets. These were slow to gain attention in the earliest days of planetary exploration, but have been making up for that recently. Their importance as relatively pristine examples of the material that formed the planets is recognised by a large section of the scientific community, who had for years to make do with meteorite samples. The satellites of the outer planets form miniature solar systems and exhibit a number of fascinating anomalies, including Saturn's giant cloud-shrouded moon Titan and the icy ocean world Europa orbiting Jupiter.

Many and varied study groups are at work more or less continuously in the space-faring nations to plan and continually refine 'road maps' for future missions to the planets. In the USA, perhaps the most important occurs each decade when NASA and its partners ask the National Research Council to look ten or more years into the future and prioritise research areas and observations, and define missions to make those observations. In 2011, the latest of these Decadal Surveys produced (amongst a lot more paper) the top-level strategy shown in Figure 16.1.

The survey team had many inputs, amongst them the 2009 report of a Venus Science and Technology Definition Team set up by NASA with the task of defining the science objectives for a possible flagship class mission to Venus that could launch in the mid-2020s. They duly came up with a concept that addresses three broad (the favoured buzzword these days is 'overarching') science goals:

1. Understand what Venus's greenhouse atmosphere can tell us about climate change.
2. Determine how active Venus is (including the interior, surface and atmosphere).

Theme	Key Questions	Missions
Building New Worlds	What initial conditions, processes and materials were involved in the formation of the Solar System? How did the giant planets and their satellites form, and achieve their present orbital positions?	Comet Sample Return Trojan Asteroid Tour Jupiter Europa Orbiter Uranus Orbiter & Probe Trojan Asteroid Tour Io Observer Saturn Probe Enceladus Orbiter
	What governed the accretion and evolution of the inner planets and their atmospheres, and what was the role of impacts in the supply of water and other volatiles?	Mars Sample Return Venus In-Situ Explorer Lunar Geophysical Network Lunar Sample Return Trojan Asteroid Tour Comet Sample Return **Venus Climate Mission**
Planetary Habitats	What were the primordial sources of organic matter, and does organic synthesis continue today?	Mars Sample Return Jupiter Europa Orbiter Uranus Orbiter & Probe Trojan Asteroid Tour **Venus Climate Mission** Enceladus Orbiter
	Did Mars or Venus host aqueous environments conducive to life? Did life emerge?	Mars Sample Return **Venus In-Situ Explorer** **Venus Climate Mission**
	Are there environments in the Solar System other than Earth that can sustain life, and is life present there now?	Mars Sample Return Jupiter Europa Orbiter Enceladus Orbiter
Solar Systems	Do the giant planets and their satellites help us to understand the formation and dynamics of the Solar System and other planetary systems? What bodies endanger and what mechanisms shield the Earth's biosystem?	Jupiter Europa Orbiter Enceladus Orbiter Saturn Probe Comet Sample Return
	How do planetary atmospheric studies lead to a better understanding of climate change on the Earth?	Mars Sample Return Jupiter Europa Orbiter Uranus Orbiter & Probe **Venus In-Situ Explorer** **Venus Climate Mission** Saturn Probe

Figure 16.1 The National Academy of Sciences recommended a road map similar to this to NASA to guide their plans for Solar System missions in the period 2013–2022.

3. Determine where and when water, which appears to have been present in the past, has gone.

Labouring (for the time being at least) under the unfortunately tedious name of the *Venus Flagship Design Reference Mission*, this ambitious programme aimed to deploy two landers, two balloons and a well-equipped orbiter, hopefully all at the same time. NASA duly noted that, setting the obvious synergisms aside, this could easily be broken up into a number of smaller and more affordable projects with objectives cherry-picked from the list. These could then compete more effectively for a new start than a single giant composite mission at a correspondingly higher price.[1]

Many of the features of the Design Reference mission made it into the final list of 26 candidates studied by the Decadal Survey teams. Encouragingly for the Venus science advocates, there were no less than four that were of direct relevance to them: *Venus Climate Mission, Venus In Situ Explorer, Venus Mobile Explorer* and *Venus Intrepid Tessera Lander*. By the end of the Decadal Survey, the last two of these had fallen by the wayside, along with most of the other 24 (the *Asteroid Interior Composition Mission* and the *Neptune System Mission* were abandoned without even completing the studies). The surface-focused missions were the result of a lot of work and aspiration by the teams who brought them forward, and they deserve a brief description before we leave them behind.

The *Tessera Lander* was to touch down safely in one of the eponymous highland regions to investigate the surface chemistry and mineralogy and conduct a photographic survey of the structure and layering of the surrounding rugged landscape. The technology requirements for getting down safely, surviving and operating sophisticated sampling and analysis instruments were clearly formidable, and explain why the proposal was deferred by the Survey team into the undetermined future.

The *Mobile Explorer* was even more ambitious. The payload would have the capability not only to land and survive, but also to move around to investigate sites several kilometres apart. It would do this not by driving across the surface Mars rover style, but using a clever lifting system, a kind of balloon in the form of a metal bellows that could expand and contract in the vertical direction to provide variable amounts of lift. Again, the technological difficulties are obvious and these, rather than lack of scientific appeal, sent the concept to the back burner, along with the Mercury Lander, Lunar Geophysical Network, Titan Lake Probe, and even the charming but ambitious Chiron orbiter.[2]

The other two Venus missions did, however, make the shortlists, one as a 'medium' mission (defined as having a likely cost of less than US$1 billion, which made them eligible for NASA's existing *New Frontiers* programme) and the other as a more expensive 'flagship' mission requiring large sums and probably politically complex international collaboration agreements. The final selection for the cheaper missions was, in alphabetical order:

- *Comet Surface Sample Return*
- *Lunar South Pole-Aitken Basin Sample Return*

---

[1] The flagship mission had an estimated price tag of around $3 billion in 2009. This is a large but not impossible cost; the current *Cassini* mission to Saturn costs something similar when mission operations over an extended period are included. Of course, there is usually a big difference (the factor being typically, some say, mysteriously equal to π) between the pre-planning estimates and the final run-out cost of any mission. Some of this is incurred deliberately, for example by extending a successful mission beyond its planned lifetime, as has happened several times with *Cassini*.

[2] Chiron is one of the larger Centaur classes of asteroids, about 220 kilometres wide, in an eccentric orbit extending well beyond, and coming just inside, that of Saturn.

- *Saturn Probe*
- *Trojan Tour and Rendezvous*
- *Venus In Situ Explorer.*

The *Venus In Situ Explorer* (*VISE*) had already been a runner-up in an earlier *New Frontiers* selection competition and was being advanced for a future opportunity, boosted by its support from the Decadal Survey. The feeling was nearly universal that studies of Venus from orbit have gone about as far as they can in making key advances. The next set of crucial objectives would require *in situ* investigations, especially to obtain the abundances of trace gases, sulphur, light stable isotopes and noble gas isotopes. *VISE* could also, it was claimed, understand the weathering environment of the crust of Venus and find evidence of past hydrological cycles, oceans, and life, and set constraints on the evolution of the atmosphere of Venus.

To convert *VISE* from a concept into a real mission, a flight proposal was submitted to the 2009 New Frontiers opportunity. With some changes and renamed *SAGE* (*Surface and Atmosphere Geochemical Explorer*), the mission was selected as one of three candidate missions for detailed technical study. However, in the final selection NASA preferred to give the nod to an asteroid sample return mission called *OSIRIS-REx*.[3]

The Decadal Survey also chose five flagship missions, which were, in order of priority with their estimated costs:

- *Mars Astrobiology Explorer-Cacher*, US$3.5 billion;
- *Jupiter Europa Orbiter*, US$4.7 billion;
- *Enceladus Orbiter*, US$1.9 billion;
- *Uranus Orbiter and P*robe, US$2.7 billion; and
- *Venus Climate Mission*, US$2.4 billion.

So Venus came last, but at least it was still in the game. With this powerful endorsement, the *Venus Climate Mission* (*VCM*) has a chance to progress towards its goals of investigating carbon dioxide greenhouse effects, dynamics and variability, surface/atmosphere exchange, and atmospheric origin and evolution. The plan includes a carrier spacecraft that deploys a mini-probe, and two drop sondes, each lasting 45 minutes as they descend to the surface, and a balloon with instrumented gondola to carry out a 21-day science campaign below the cloud tops.

The proponents of *VCM* pointed out that the mission 'will return a dataset on Venus's cloud properties and radiation balance and their relationships and feedbacks, which are among the most vexing problems limiting the forecasting capability of terrestrial GCMs. Evidence will also be gathered for the existence, nature and timing of the suspected ancient radical global change from habitable, Earthlike conditions to the current hostile runaway greenhouse climate, with important implications for understanding the stability of climate and our ability to predict and model climate change on Earth and extra-solar terrestrial planets. This mission does not require extensive technology

---

[3] OSIRIS-REx stands for Origins Spectral Interpretation Resource Identification Security Regolith Explorer. No comment.

development, and could be accomplished in the coming decade, providing extremely valuable data to improve our understanding of climate on the terrestrial planets.'

Venus in NASA's *Discovery* programme

In the early 1990s, NASA was in the grips of Administrator Daniel Goldin's 'faster, better, cheaper' philosophy for planetary missions, and the *Discovery* programme was to be its way forward. Missions would be selected in a competition, rather than by programme committees, and had to satisfy a very low cost cap. They would be proposed and led by a scientist managing a team drawn from industry, small businesses, government laboratories, and universities. High risk was to be tolerated, although it turned out later that this was only true if they didn't fail.

Discovery was a great opportunity for Venus missions, since the planet was relatively easy to reach with small spacecraft on cheap launchers. At this time, the top objectives (particularly the processes producing the hot surface, and driving the super-rotating atmosphere) could be addressed with fairly simple orbiters and entry probes at a cost that was not too difficult to keep under Mr Goldin's cost cap, which at first was only $200 million.

Priorities and road maps: NASA leaves Venus on the back burner

In February 1995 NASA selected the *Venus Multiprobe Mission* (*VMPM*) for the shortlist to receive funds for a detailed study. Essentially an expanded and improved version of the *Pioneer Venus Multiprobe* mission of 1979, *VMPM* would drop no less than 16 probes this time, to try to nail down the elusive characteristics of the atmospheric circulation. The probes would measure winds, temperatures and pressures from an altitude of 60 kilometres down to the surface, with better precision, as well as coverage, than *Pioneer Venus*. The theoreticians who planned to compare the data with computer models of the circulation also wanted high vertical resolution to reveal the wave structures propagating angular momentum upwards from the surface. *VMPM* could deliver measurements spaced as finely as just 10 metres apart near the bottom of the atmosphere, where the probes would be moving slowly after shedding most of their entry velocity in the upper atmosphere.

Wind measurements would be obtained by tracking the probes using dual-frequency, differential, long-baseline interferometry, with four receiving stations on the Earth. The precision of the wind data needed to be at least ten times better than those from *Pioneer Venus*, again to resolve the transient eddy circulations; better than 5 centimetres per second was anticipated, about 1 per cent of the mean meridional wind. The team would develop its own numerical atmospheric general circulation models containing all known physical processes and use them to distinguish different dynamical elements existing in the data. Sadly for Venus, in the final judgement *VMPM* lost out to the *Stardust* mission, which flew in 1999 to return dust samples to Earth from comet Tempel-2 in 2005.

At about the same time, *Venus Environmental Satellite* (*VESAT*) was being developed as an inexpensive alternative approach to addressing the circulation mystery with remote-sensing instruments from orbit. *VESAT* was to acquire three-dimensional global maps of winds, temperature fields and trace gas abundances from a 45-degree inclined, 30,000-kilometre altitude circular orbit. Although the spacecraft can be inexpensive, orbiters tend to have high mission operation costs because of their long lifetime. The *VESAT* team designed a plan for spacecraft operations that would achieve coverage of the entire globe with minimal day-to-day intervention from the mission controllers, who would be located in an inexpensive university setting rather than the usual NASA centre.

A single contractor, Ball Aerospace in Colorado, was to be responsible for the design, manufacture and integration of the entire *VESAT* spacecraft, including the instrument payload. Again this was a major departure from common practice. Usually, each responsible scientist produced the instrument that addressed his or her key objectives, and delivered them in a flightworthy state for later integration by NASA. The *VESAT* team claimed that significant savings in cost and schedule would accrue from the integrated approach. All was in vain, however, perhaps mainly because the proposal had been upstaged by the selection for flight in Europe of *Venus Express*, a rather similar mission.

The *VESAT* team turned its attention to *in situ* measurements from aerostat balloons and came up with *Venus Atmospheric Long-duration Observatories for in situ Research* (and its stirring acronym, *VALOR*). Like *VESAT*, this addressed questions about the planet's origin, evolution, chemistry and dynamics as identified in the Decadal Survey, but would trade global coverage for direct sampling of atmospheric temperature and pressure, cloud particle sizes and their local column abundances, the vertical wind component and the chemical composition of cloud-forming trace gases. *VALOR* was proposed in the *Discovery* competition of 2004, and again in 2006, but lost both times.

VALOR was handicapped by the need to deploy its balloons at latitudes near the equator, where abundant solar power is available, whereas observations from orbit have shown that much of the interesting atmospheric dynamics on Venus occurs near the poles. Recently, work has been going on to produce a nuclear-powered version of the aerostat that would overcome this limitation. It would deploy at the highest latitude easily reachable with a modest launch vehicle, and ride the winds at an altitude of 55 kilometres while drifting polewards. Powered by a Stirling radioisotope generator, an advanced version of the radioisotope thermal generators already used on missions to the outer Solar System such as the *Cassini* Saturn orbiter, and much lighter and more efficient, the super-*VALOR* would have a longer lifetime (partly because of better thermal control using the extra power available) and improved instrumentation, as well as a more fascinating location to explore.

Exciting as it is to contemplate a large instrumented balloon cruising above the cloud tops near the pole and investigating the dynamics, meteorology and chemistry in the wonderfully complex polar vortices, the present reality is that the advanced power sources are not yet available. (The team makes lemonade from this by pointing out that *Polar VALOR* would be a great way to space qualify the devices – which clearly have many other applications – through all mission phases and in various operating environments.)

All was not lost for Venus and *Discovery*; in October 2006 NASA selected a mission called the *VEnus Sounder for Planetary ExploRation* (leading to another nifty acronym, *VESPER*). It was never completely clear, even to those of us involved in the process, why NASA did not see *VESPER* eclipsed by *Venus Express*, as *VESAT* had been; all three fit the *VESPER* description as 'a Venus chemistry and dynamics orbiter that would advance our knowledge of the planet's atmospheric composition and dynamics ... [by] the first comprehensive and synoptic study of Venus with sufficient sensitivity and duration to test major models of the dynamics, chemistry and circulation throughout the Venus atmosphere'. *VESPER* uses some different and exciting techniques: the primary instrument would be the *Submillimeter Limb Sounder* (SLS), capable of high spectral resolution for the sensitive detection of trace gases and measurement of Doppler wind velocities above the clouds. Infrared and ultraviolet imagers would track the dynamics of the lower and upper clouds, and X-band radio occultation would provide high spatial resolution temperature profiles.

Although it got further down the track than *VESAT* or *VALOR*, *VESPER* too fell by the wayside eventually. NASA has selected a dozen *Discovery* missions since 1992, headed all over the Solar System – except to Venus.[4]

The Russian revival

Since the break-up of the Soviet Union, the Russians have maintained an active launch capability – now commercialised, and paid to dispatch, for instance, *Venus Express* for the European Space Agency (ESA) – but have been slow to get back into the business of sending their own planetary probes. After the failure to get *Phobos Grunt* to Mars in 2011, the Russian Federal Space Agency has plans for another Mars mission (*Mars-NET*) and two to the Moon, plus a new mission to Venus, *Venera-D*, which the government had approved in 2005 for flight in 2013 (Plate 21).

There is still something of a feeling, and not just in Russia itself, that Venus is a 'Russian planet' where space exploration is concerned. This reputation was well earned by the ambitious and successful programme that was mounted in the 1960s, 1970s and 1980s with the *Venera* and *Vega* missions. *Venera-D* was meant to continue this tradition, with a massively comprehensive list of objectives requiring several spacecraft and considerable new technology, including advanced balloon platforms known as aerostats.

Venera-D took various forms, but at its most ambitious it would involve the emplacement of two orbiters, four balloon-borne instrumented packages in the clouds, a lander

---

[4] The latest, in 2011, was Interior Exploration using Seismic Investigations, Geodesy and Heat Transport (InSight), a Mars drilling mission. Although, with *VESPER*, this is a candidate for the prize for the clumsiest name that condenses to a smart acronym, the winner has to be the 1997 *Discovery* winner, Extrasolar Planet Observation and Deep Impact Extended Investigation (EPOXI).

on the surface and a number of deep-atmosphere probes (Plate 22). The probes would be small, and intended to provide ballast that could be shed by the balloons to prolong their lifetime at night, when they would otherwise tend to sink, as well as to make measurements in the relatively inaccessible lower atmosphere. One of the two orbiters was a radar mapping satellite similar to *Venera 15/16* but with higher resolution on the surface, and the other a satellite in a different orbit dedicated to remote sensing of the composition of the atmosphere and its circulation patterns. A pair of balloons would be released from the lander at two different altitudes during its descent to float near the cloud tops and measure acoustic and electrical activity for up to eight days. During this period the balloons would drop four microprobes at different locations for a 30-minute descent to probe the atmosphere down to the surface. Finally, the possibility of a kite-like glider to ride the winds at an altitude just below the cloud base for at least a month was seriously considered.

Since its inception *Venera-D* has trod a tortuous path of modifications and delays that would also be familiar to NASA and ESA. The 'D' stands for *Dolgozhivushaya*, which means long duration, referring to the part that will land and make measurements on the Venusian surface. Long is relative, of course; on 19 May 2005, the Space Council of the Russian Academy of Sciences held a meeting at which it considered how long it would take to carry out the key experiments, such as the seismological measurements to detect 'venusquakes'. The time allotted has to be traded off against the obvious difficulties. Finally, they defined long duration as 30 days, to give the engineers something to work towards. Since the record up until now is just over an hour, this would be a considerable advance, not to mention a formidable technical challenge.

The mission would have been ambitious even without the problem of keeping instruments designed to analyse the soil beneath the lander working for a month when the temperature outside is over 400 degrees centigrade. The orbiting component of the mission, in addition to pursuing its own research objectives, would need to relay the data back to Earth from the payloads in the atmosphere and on the surface of Venus. This is much more difficult than it sounds. If the data rates are to be sufficient for imaging, spectroscopy and high-resolution sampling by mass spectrometers, as the latest scientific goals would require, then a suitable combination of storage media on the *in situ* platforms and frequent line-of-sight contacts with the orbiting relay spacecraft would have to be designed. High performance antennae, preferably pointable, would have to be provided at both ends.

Not surprisingly, the mission was gradually simplified, or 'descoped' as NASA would say. The radar mapping satellite was dropped, and a report from 24 January 2007 gave the revised scientific goals as follows:

Orbiter
- The 3D temperature and wind field in the middle atmosphere, its local time and temporal variations. Investigations of the thermal tides;
- Nature, composition and optical properties of the clouds;
- Nature of 'unknown' ultraviolet absorber;
- Dynamics and nature of super-rotation;
- Chemical composition of the atmosphere: H_2O, SO_2, CO, HCl, HF, etc.;

- Surface temperature, volcanic activity, lightning;
- Plasma environment.

Balloons

- Dynamics of the atmosphere;
- Meteorology (pressure, temperature density with high accuracy);
- Chemical composition and optical properties of the clouds;
- Chemical composition of the atmosphere;
- Imaging of the surface;
- Radiative balance and greenhouse effect;
- Surface temperature;
- Possible volcanic activity, lightning.

Lander

- Chemical composition of the lower atmosphere;
- Abundance (with high accuracy) of noble gases and isotopic composition;
- Clouds composition and optical properties;
- Radiative balance and greenhouse effect;
- Surface temperature, mineralogical characteristics;
- Possible volcanic and seismic activity, lightning.

By 2011, many of the capabilities of the mission had been further downgraded and the launch date delayed to 2017. Attempts were made to achieve the desired lifetime by devising ways to study the atmospheric composition and clouds during descent, and to analyse surface material after landing, both without bringing samples inside the lander. Eventually, however, the plan to operate for a month on the surface was deferred to a later, more hypothetical mission called *Venus-Glob* which might fly in 2021 or beyond. The lifetime problem remains unsolved, but the Russian scientists say they are not giving up.

European plans

ESA differs from NASA in that suitably qualified groups of scientists from all parts of the European community can respond to 'Announcements of Opportunity' for complete missions, as well as for individual experiments on those missions. In 1999 a French-led consortium from eight different nations collaborated on a proposal for a Venus probe mission which was given the name of *Lavoisier*.[5] The opportunity was for a mission of modest cost, so the plan was to release a probe and three balloons but not to go into orbit, limiting the data-taking phase for the balloons to 24 hours. In this sense the project was a pathfinder for longer-lived missions using entry techniques, rather than the last word on Venus entry science. However, with the payload focused on chemistry rather than dynamics, it would still be able to measure the crucial noble gas elemental and isotopic composition, the chemical cycles in the clouds, and the surface–atmosphere interactions between gases and minerals.

---

[5] After Antoine Lavoisier, the brilliant French chemist, a nobleman who was guillotined during the revolution.

Lavoisier lost out on that occasion to an astronomy project but did seem to instil a belief in ESA's committees that taking on a mission like this was only a matter of time. This can, and in this case did, lead to expenditure on several 'Technology Reference Studies' of possible new missions with Venus as the target. Such studies are carried out regularly on all sorts of promising but immature project concepts, often with no intention that any of them will fly any time soon. Instead, the idea is to start the development of new technologies that are likely to be required eventually, early enough that the mission is a serious and practical option for future selection in competition with other areas of science.

For Venus, the agency decided that an entry probe would be a logical successor to the *Venus Express* orbiter, since it is technically more challenging and opens up new areas of science that would be complementary to the achievements of the earlier mission. The objectives would have to be an advance on the successful *Venera* atmospheric and landing probes deployed between 1967 and 1981, the *Pioneer Venus* probes (1978) and the *VEGA* balloons (1985), but this is not too difficult in view of the technological advances that have taken place in the past 20 years, and the refocusing of objectives in the light of results from *Magellan* and *Venus Express*.

The objectives for a European probe mission, unshackled for the moment from too many cost constraints as no hardware was involved, were, as we might expect, not too different from the corresponding NASA mission studies. The ESA study team, with many of the *Lavoisier* proposers involved, highlighted how the origin and evolution of the atmosphere can be investigated by *in situ* measurements of the isotopic ratios of the noble gases, and how accurate measurements of minor atmospheric constituents, particularly water vapour, sulphur dioxide and other sulphur compounds, will improve our knowledge of the greenhouse effect on Venus, atmospheric chemical processes and atmosphere–surface chemistry, and will address the issue of the possible existence of volcanism.

To make significant advances on past measurements would require sophisticated instruments, including an advanced mass spectrometer (Figure 16.2). The cause of temporal and spatial variations of the cloud layer opacity and measurements of the size distribution, temporal and spatial variability as well as the chemical composition of the cloud particles were on the list. So too was a somewhat gratuitous remark, carefully phrased: 'Furthermore, it has been suggested that the unidentified large (about 7 microns in diameter) cloud particles might contain microbial life.' The team did not think this sufficiently likely to include life detection instruments in their model payload.

A single probe can do relatively little about objectives related to the general circulation and super-rotation, the complex magnetosphere, or the surface geology and tectonics. Recognising this, ESA's 2005 study added dual orbiters, buoyant stations and multiple small probes ('drop sondes') to a study that looked towards a launch of the whole flotilla in 2013 (sadly this is not going ahead, at least not yet). It also seems likely that a successful probe will be followed by a soft lander, and possibly even sample return. Although all of these have been studied by ESA, and research is going on at various levels into the technology they would require, the agency and the community representatives on its

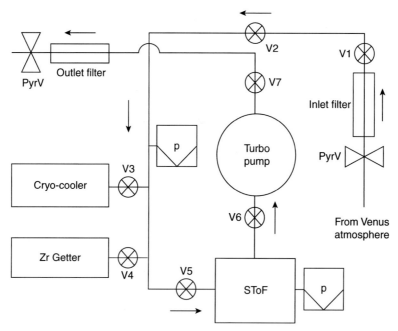

Figure 16.2 This schematic of the mass spectrometer proposed for the *Lavoisier* balloon payload gives some idea of the complexity of the plumbing required. The instrument, itself a very complex piece of kit, is in the box labelled 'SToF'.[6]

committees have shown little inclination to implement an actual new mission to Venus, to follow on from the success of *Venus Express*.[7]

Part of the reason, of course, is that the old mission is still going, and until it is over at the end of 2014 or thereabouts, the impetus to spend large sums on a successor is lacking. Of course, it would make perfect sense to start now on the new project, rather than planning a gap, but that is not how it works unfortunately. Too many other objectives are in the queue for too little funding.

But the studies are being taken seriously and the route to the new hardware and the scientific support clearly exists, so that these new missions will very probably be implemented eventually, probably in much the form now being studied. We will therefore look at them in more depth in Chapter 17, where we review what are likely to be the realities of Venus exploration in the next few decades.

---

[6] SToF stands for spiral time-of-flight. In this type of mass spectrometer, the incoming atoms and molecules from the atmosphere are ionised so that they carry a positive charge, and can be deflected by a magnetic field inside the instrument. The deflection is different for particles of different mass, so they become separated and their abundances can be measured. The 'spiral' part refers to the path the ions follow: early mass spectrometers just bent the particle path, but better results can be obtained with a spiral because the total distance travelled and hence the separation is greater.

[7] As a multinational organisation, ESA's committee structure has political as well as scientific elements. The top-level scientific body is the Science Programme Committee, which has under it advisory groups for Space Science, Astronomy and Solar System research.

Chapter 17
Coming soon to a planet near you: planned Venus missions

As a result of the labyrinthine processes described in Chapter 16, we arrive at a 'best guess' for the near-term (next two decades, say) future of Venus exploration in the form of some combination of entry probes, balloons and landers. These will come from NASA and the European Space Agency, possibly in tandem but more likely not; and *Venera-D* from the Russians. Japan may try again to orbit Venus, and something from the Chinese and Indians cannot be ruled out, although they are more likely to focus on the Moon and Mars.

Such a programme is by no means assured, of course; there could be no new mission to Venus for 20 years, at the end of which time everything will have changed. It would be nice to think that several of the world's space agencies might get together and pool their resources in the future, to mount a single large mission, perhaps sample return. However, history suggests a more fragmented approach can be expected, at best. Despite all of the uncertainty, we now look, in a spirit of optimism tempered with realism, more closely at the plans as they stand and will likely evolve in this possible multi-pronged attack on the remaining mysteries of Venus.

The next NASA mission to Venus

The proposed *Surface and Atmosphere Geochemical Explorer (SAGE)* mission to Venus is, it will be recalled from Chapter 16, similar to the *Venus In Situ Explorer (VISE)* concept which preceded it (Plate 23). The defining characteristic of the project is a large probe (Figure 17.1) that will descend through the planet's atmosphere, making measurements of the composition and obtaining meteorological data. The focus would be on what the noble gases and their isotopes reveal about evolution, but it would also obtain trace gas profiles, especially for the sulphur compounds, for cloud-related chemical cycles and surface–atmosphere interactions including volcanism and weathering.

The probe would land on the surface of Venus, where it would survive rather longer than the old *Veneras* did, giving it time to find and expose, using an abrading tool, the most interesting areas of rock and soil it can reach (without moving), and measure their composition and mineralogy at and below the surface. At the same time, other instruments would measure the atmospheric composition near the surface in unprecedented

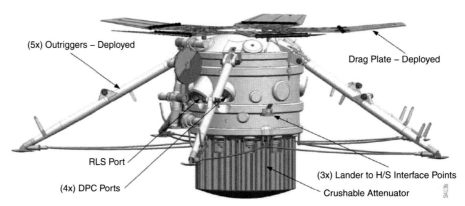

Figure 17.1 The *Surface and Atmosphere Geochemical Explorer* (*SAGE*) concept. The drag plates replace parachutes for slowing the descent in the dense lower atmosphere, and the crushable module on the base absorbs the remaining impact while the outriggers keep the housing upright. The apertures shown are for the descent panoramic camera (DPC) and the Raman laser spectrometer (RLS).

detail. Thus *SAGE* will tell us about the surface weathering on Venus, and maybe even whether there was a liquid water ocean in the past. It would also take more and much better pictures of the terrain, both during descent and after landing, than the few we have now from the Russian landers, taken more than two decades ago.

With a bit of luck, we will also see from NASA in the next decade or so, some version of the *Venus Climate Mission*. This will target the key objectives that a single lander cannot, namely the global circulation/super-rotation question, and the detailed nature of the greenhouse climate of Venus, with its promise of advances in the understanding of climate stability and global change on Earthlike planets. Unlike the static, one-off surface investigation, which of course has plenty of its own technical challenges, understanding climate dynamics requires three-dimensional, time-varying data on radiation balance, atmospheric motions, cloud physics and atmospheric chemistry and composition of the middle and upper atmospheres in order to identify the fundamental climate drivers.

Some of these goals need planetwide coverage from an orbiter, others long-duration, *in situ* sampling from an aerostatic station, and finally deep-atmosphere probes to sample the depths that orbiters and balloons have trouble observing in sufficient detail. However, the cost and complexity of doing all of these things at once tends to lead to non-selection. The *Climate Mission* proposers compromised by replacing the orbiter with a flyby carrier spacecraft that would deliver the *in situ* package but not itself stay at Venus, thus saving the very considerable mass of the orbit injection motor and its fuel. Orbiter instruments and operational costs such as tracking and data downlink would also be unnecessary if it could be argued successfully that missions such as *Venus Express* had already addressed the key science possible from orbit, which is true up to a point.

Even so, the launch mass of the carrier, aerobot and three small probes works out at just under 4 tons, a huge mass to take to Venus, but possible with an *Atlas V* launcher and a five-month flight time. Current plans call for a launch on 11 November 2021. One of the probes is larger than the other two, the latter being tiny drop sondes, similar to those developed by the Europeans and discussed in the next section. The larger probe, still quite small with a mass of less than 40 kilograms, carries a mass spectrometer for composition measurements as well as a net flux radiometer and temperature and density sensors. The sondes, weighing in at 12 kilograms, have only the radiometer and sensors.

The balloon (Plate 24) has to be inflated inside Venus's atmosphere, leading to quite a complex entry sequence (shown graphically in Figure 17.2) to ensure the station becomes self-supporting before the whole package descends to the intolerably hot regions below the clouds. A study by NASA's Ames Research Center showed that it need take no longer than five minutes, at which point the altitude is 53 kilometres above the surface. When

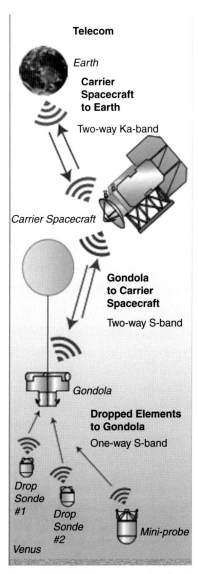

Figure 17.2 NASA's concept for the *Venus Climate Mission* and its deployment sequence.

fully inflated, the balloon jettisons the helium cylinders and releases the probe, rising to its float altitude of 55 kilometres as a result. Once there, it is designed to survive for 21 days before its batteries run out of power. This is long enough, starting at a mid-southern latitude, for the payload to sample the polar regions, spiralling around the planet five times in the rapid zonal winds before reaching the edge of the giant polar vortex, after which its path becomes fascinatingly unpredictable as it approaches the maelstrom at the core of the 'dipole'. The two drop sondes would be released on command somewhere along this trajectory, the choice of location depending not just on science but also on whether the aerostat needs to shed ballast weight. Hopefully, at least one of them can be saved to probe the depths in the unusual conditions near the pole for the first time.

European Venus mission studies

The European Space Agency already has experience of entry-probe missions to planets with the successful *Huygens* landing on Titan in 2005. Since then, work has been done on arrival, entry and descent mission profiles to define velocities and communication windows for a Venus probe either on a direct hyperbolic trajectory or carried as a piggy-back on board a larger mission that uses a Venus gravity assist on the way to one of the outer planets.

ESA's thinking is more modest in mass terms than NASA's. The European probe mass target was less than 300 kilograms, with only about 10 kilograms of scientific payload, assumed to be a camera and a mass spectrometer. This compares with around 650 kilograms for the entry component of the NASA mission.

A *Soyuz-Fregat 2–1B* rocket launched from Kourou, similar to that used for *Venus Express*, can send a total of about 1,400 kilograms to Venus, of which about a ton ends up in orbit after fuel has been burned for orbit insertion. In a 2005 study, ESA looked at using this to deploy simultaneously a pair of small satellites and an aerobot, the latter carrying no fewer than 15 lower-atmosphere microprobes.

One satellite will be in a polar orbit, carrying a remote sensing payload suite primarily dedicated to making context measurements in support of the *in situ* atmospheric sampling by the aerobot. The second satellite enters a highly elliptical orbit, deploys the aerobot and subsequently operates as a data relay satellite, while it also performs science investigations of the ionosphere and radar mapping of the surface. The aerobot consists of a long-duration balloon, smaller than the American one but equipped for investigating the cloud layer chemistry. The microprobes are smaller but more numerous than in the American plan, each capable of determining the vertical profiles of pressure, temperature, flux levels and wind velocity in the lower atmosphere.

Like *Venus Express*, at launch the *Soyuz-Fregat* sends the payload into a highly elliptical Earth orbit, then the upper stage fires again providing the impetus for Earth escape. The journey takes the spacecraft halfway around the Sun, spiralling inwards, until Venus is reached in about 150 days. The three-axis stabilised polar orbiter has a mass budget of 30 kilograms for the remote-sensing atmospheric science instruments. This assumed the concept, fashionable in ESA at the time, of a highly integrated payload suite, which merged individual instruments onto one platform thereby (on paper at least) achieving large mass and power reductions without sacrificing the scientific performance. It was pretty obvious to those of us who had built instruments and were working with ESA at the time that the idea was too optimistic, or too far ahead of its time, and it became the European equivalent of the savings NASA expected from its 'faster, better, cheaper' concept, ending up in the bin.

The polar orbiter has no communications link to the Earth, relying instead on relaying the science data to the elliptical orbiter through an X-band link. The latter stays in its highly elliptical orbit until after it releases the entry probe, also acting as a relay station for transmitting data from the aerobot to Earth. After the operational phase of the aerobot has ended, the elliptical orbiter will progress to a lower orbit suitable for operating the ground-penetrating radar and the radar altimeter.

The spherical storage tank for the gas needed to inflate the balloon takes up most of the volume of the entry probe. New developments, such as chemical gas generators or solid-state storage of hydrogen (potentially important for hydrogen-powered cars on Earth) could save a lot of weight and volume. The probe is shaped to be stable in the hypersonic and supersonic regimes, so that no active control is required. It enters the atmosphere with a velocity of just under 10 kilometres per second and a steep flight path angle, intended to keep the entry sequence as short as possible and ensure a quick release of the aeroshell. This will minimise the time available for the heat generated to soak through the heat shield into the lander. While still travelling at supersonic speed, a parachute is deployed by a pyrotechnic device to stabilise the probe as it decelerates through the sound barrier. The front aeroshell is released as soon as the subsonic regime has been reached, a few seconds after parachute deployment. At a velocity of 20 metres per second the balloon can be unpacked and inflation started. The gas storage system is released when the balloon is inflated, and the aerobot gradually finds its cruise altitude, stabilising at a height of 55 kilometres above the surface. At this altitude the scientific investigation can be addressed in a relatively benign environment (a temperature of 30 degrees centigrade and half an atmosphere of pressure), where it will hopefully stay stable for long enough to travel around Venus at least twice.

The *Vega* balloons were carried by the wind at about 70 metres per second at a similar altitude. If this performance is repeated by the new balloon, the flight needs to last for 14 days to achieve the mission objective. As the gas in the super pressure balloon is heated by the Sun, the float altitude will increase and gas will gradually escape. Eventually there is insufficient gas left for positive buoyancy at night, and the balloon sinks and finally fails. Carefully planned microprobe drops could partially compensate for the loss of gas and maximise the operational lifetime, but it would be better to have a gas release and replenishment system to handle the diurnal cycle. This would allow an operational life-time of about 30 days. Hydrogen would probably be used for the balloon inflation gas, with helium as a backup because, although it is safer, as a monatomic gas it is more prone to leakage.

Figure 17.3 shows a conceptual drawing of the gondola layout and its suite of scientific instruments. A gas chromatograph/gas spectrometer combination performs composition measurements, a nephelometer is used to obtain cloud density profiles, solar and infrared flux radiometers provide the radiative heating and cooling versus height, a meteorological

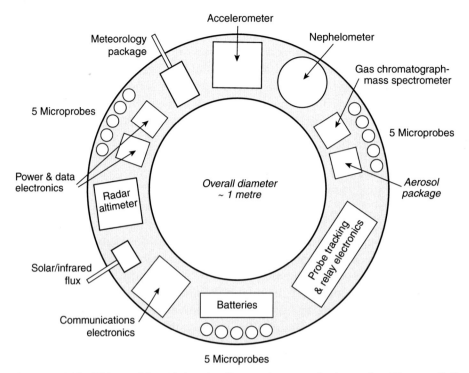

Figure 17.3 The ESA gondola with its scientific experiments and microprobes. The overall diameter is about a metre, and the claimed total mass just 4 kilograms.

package delivers temperature and pressure data, a radar altimeter studies the topography below, and the microprobes and their deployment system make up the rest of the payload.

The consultants advocating and designing the highly integrated payload suite came up with a total mass of 4 kilograms and an average power consumption of 5 watts. As already noted, some parts of ESA had undergone a Damascene conversion to incredible levels of integration and miniaturisation at around this time, and both of these numbers are probably about an order of magnitude too small in practice, but if achievable they would allow silicon solar cells mounted on the gondola surfaces to provide sufficient power during the day, with batteries for night time. The entire gondola weighs in (in the study) at only 23 kilograms, compared with more than ten times as much for the NASA version.

The 15 microprobes, with a mass of only 100 grams each,[1] contain a small payload that will measure pressure, temperature and solar flux levels from the aerobot float level down to an altitude of 10 kilometres or deeper (Figure 17.4). The horizontal wind velocity can be deduced by careful tracking of the trajectory of the microprobes by the gondola.[2] The

[1] About the same as an iPhone 5.

[2] A feasibility study, 'DALOMIS', by QinetiQ in England came up with a strategy for tracking several probes simultaneously with the desired accuracy.

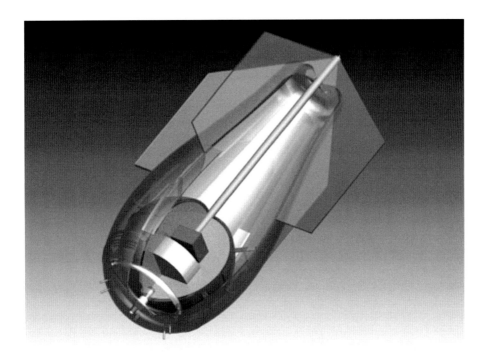

Figure 17.4 A European concept for a Venus microprobe. The casing is made of high-temperature-resisting glass, and contains temperature and pressure sensors, upwards and downwards light flux meters and communication/tracking electronics. Designed to be deployed in a small swarm, each probe is 11 centimetres long and its total mass is 100 grams.

strategy for dropping the probes is a compromise between the desire for coverage in space and time, especially in latitude to characterise the global atmospheric dynamics; for small 'swarms' to measure localised gradients especially in winds, to detect turbulence and investigate the local weather patterns on Venus; and finally the need to shed mass when the balloon loses lift through leakage of gas. The compromise was to release the 15 microprobes in five separate drop campaigns, spaced equally over the mission lifetime, with the three probes in a drop campaign released five minutes apart.

Venera-D

As we saw in Chapter 16, *Venera D* has undergone several changes during a long and difficult gestation as Russian space scientists struggle with their ambitions against a background of difficult new political and fiscal conditions in their country. It is not easy to guess what might finally fly, if anything. The soothsayer's task is not made any easier by the language barrier and some residual reticence on the part of the Russians to try not to give too much away, a hangover from the Soviet era. Then they operated under rules that basically forbad discussion of missions until after they had performed successfully, presumably in case they failed and made the administration look bad. Nowadays,

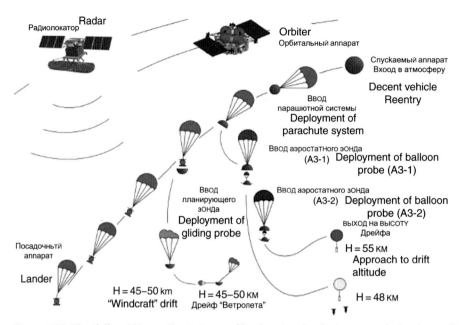

Figure 17.5 The 'full-up' *Venera-D* mission profile, showing the deployment of a lander, a glider, two balloons each carrying four dropsondes, and two orbiters, one a radar mapper and the other an atmospheric remote-sensing platform.

presentations about concepts like *Venera-D* are made at international meetings and the Russians actually host forward-looking sessions about Venus exploration and other topics in their own country.

It is still a fairly confusing picture. Basically, the Russian scientists do not themselves have a very clear vision of where their programme might be heading; approvals are given and modified or withdrawn with relatively short notice so long-term plans are even more tentative than they are in the West. But it is clear that the strategy of the Russian planetary scientists has shifted from the ambitious long-lived lander to something simpler, on the grounds that even a mission that essentially repeats something that has been done before is good science when investigating a complex world, and that doing something is better than nothing when trying to restart a once very successful programme.

Like the other nations, Russia still hopes that its next Venus mission will have several elements working at once. One of the options presented in November 2009 at a workshop in Moscow is shown in Figure 17.5, which has two orbiters, one for remote-sensing experiments and one carrying a large synthetic aperture radar, plus a lander, a glider and two balloons.

Perhaps the most exciting aspect of the plan is the simultaneous deployment of three airborne platforms at different altitudes. Two balloons will float for at least eight days at different heights inside the clouds, and the glider or 'windcraft' at 45 kilometres or possibly lower. Thus they will span the vertical extent of the main cloud layer, and obtain simultaneous wind speeds at three heights, all enticing prospects after years of trying to understand complex time-dependent phenomena from single 'snapshots' of data. The glider is intended to have the capability to swoop under the clouds and obtain the first aerial survey of the surface in normal visible light.

The orbiter carries 80 kilograms of instruments, a large mass allowing a very comprehensive set including thermal infrared, near-infrared, ultraviolet, microwave and submillimetre spectrometers that span virtually the entire electromagnetic spectrum. Within the atmosphere, the balloons carry advanced nephelometers and laser probes for analysing the clouds, along with the usual temperature, pressure, and wind tracking experiments. The lower-flying station has a mass spectrometer for atmospheric composition and the other carries microprobe drop sondes, planned to be intermediate in size between the American and European concepts with a mass of about 2 kilograms each.

This 'full-up' version requires the large *Proton* launcher and high levels of funding (the precise amount is not revealed, of course, and in any case very hard to calculate under the Russian system). At the 2009 workshop, it was announced that *Venera-D* was approved for a 2016 launch, with arrival in May 2017. Recent updates suggest this has been delayed and that it is more realistic to expect a reduced version of the mission launched on a *Soyuz* rocket in the next ten years.[3] The balloons, microprobes and glider were shelved in one version, leaving just the orbiter and a lander, with payloads very much like updated versions of the *Venera* missions of the 1980s.

The *Venera-D* orbiter would still carry several spectrometers in the spectral range from UV to millimetre, including mapping spectrometers, a camera and a plasma package. The lander would have a gas chromatograph-mass spectrometer; a meteorological package measuring pressure, temperature and wind, a nephelometer and a particle size spectrometer; gamma-ray, Mossbauer, tuneable diode laser and laser-induced plasma spectrometers; and a TV package containing panoramic, high resolution and descent cameras.

The lander will no longer be the long-lived development originally desired, but more like the past *Veneras* which use insulation to obtain an hour or so of data collection on the surface before the high temperature leads to failure. There have been considerable advances in instrumentation in the recent decades, so we can expect better measurements of the surface composition, as well as new views of a different location. The orbiter likewise will do better than its predecessors because of the advances made in the meantime, not just in hardware but in understanding where to look (in wavelength as well as location and height) for new clues to the key problems.

International cooperation

It cannot have escaped the notice of even a casual reader of this chapter that the near-term future plans of the Americans and the Europeans, and to a lesser extent of the Russians and the Japanese, are quite similar. Could they not get more done, or save a lot of money, or both, by working together? Of course they could, but it is not the simple matter it might first appear of taking the estimated cost of a mission and dividing it by two, or four, so that each partner pays a fraction.

---

[3] The *Proton* can deliver over 2 tons into Venus orbit; the smaller *Soyuz-2* about half of this.

The partnership itself exacts a considerable price in meetings, travel, translations, interface documents, legal agreements and a thousand other things that might be bundled together as 'inefficiencies'. Not only that, but space technology is often strategically sensitive and there can be an unwillingness to share (in the case of the United States, with its 'ITAR' regulations,[4] all sorts of restrictions, sometimes farcical, are enshrined into law). Finally, there is the possibility, always present, that a partner will pull out at some stage for any of a number of reasons.

The main casualty of attempts to share the cost of expensive robotic and eventually manned missions has been the exploration of Mars, when, in the aftermath of the fall of the Soviet Union the Russians and Americans developed short-lived plans to travel to the red planet together. Very recently, NASA pulled out of an ambitious joint mission to the Jovian system, leaving their European ex-partners in the lurch. ESA, strongly committed to the mission, is currently trying to work out how to carry on alone with a stripped-down version known as *Jupiter Icy Moons Explorer* or (horrors) *JUICE*. To be fair, the joint US–European mission *Cassini-Huygens*, a Saturn orbiter with a Titan probe, has been a great success, after some early panics. Similar collaborations, successful or abortive, have yet to be attempted for Venus, and there is no real sign of any happening in the foreseeable future. The whole game may change before long, however, when the emerging space nations, especially India and China, who have mentioned Venus in their plans, join in.

---

[4] ITAR sands for International Traffic in Arms Regulations, but 'Arms' are defined to include space technology and scientific instrumentation, along with associated documentation and slides shown at meetings, as well as computer software.

Chapter 18
Towards the horizon: advanced technology

Larger, cheaper payloads: aerobraking and aerocapture

In the slightly longer term, missions to Venus are likely to benefit from various technical improvements that have been made or are in an advanced stage of development. For instance, there are two new techniques that will make it much cheaper to place a substantial mass in Venus orbit without using a very large and expensive launch vehicle. Both involve interaction with the atmosphere to slow the spacecraft down on arrival, thus eliminating some of the heavy rocket motors and the fuel that would otherwise be required.

Aerocapture is the approach in which the spacecraft is targeted accurately to the right pressure level so that atmospheric drag reduces its velocity relative to the planet to the point where it is captured into orbit. There is a risk, of course, that erroneous targeting, or poor knowledge of the atmospheric conditions, can result in the spacecraft crashing into the planet, or missing it altogether. In either case the margin available is not great.

The second, related technique is *aerobraking*, where the spacecraft achieves initial orbit by conventional means, that is, using a retro-motor, and the orbit is subsequently adjusted by allowing small amounts of drag to occur by skimming through the upper atmosphere when the spacecraft is near its closest approach to the planet. By gradually lowering the periapsis over an extended period of time, aerobraking can be gradually increased and the risk is thereby reduced to an easily controlled level. Aerobraking was used successfully at Mars, most recently for the *Mars Reconnaissance Orbiter* in 2006, and at Venus by *Magellan* in 1993 (Figure 18.1).

Aerobraking at Venus with *Magellan* required some complex manoeuvres. For the drag pass, the spacecraft velocity was aligned perpendicular to the plane of the solar panels to present the maximum cross-section. After that, it turned to point the communications antenna towards the Earth to relay engineering data gathered during the passage through Venus's upper atmosphere. Next, maps of the star field were obtained to ascertain the details of the new orbit, and small trims made using the attitude control jets. For thermal control reasons, it was also sometimes necessary for *Magellan* to point the main communication dish towards the Sun, to shade the rest of the spacecraft.

Some sort of heat shield for the spacecraft is essential when frictional braking is used, although the requirement is much more demanding for capture, where large amounts

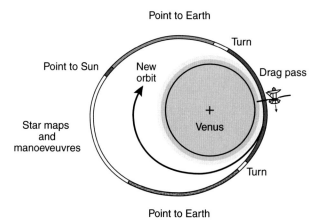

Figure 18.1 *Magellan* aerobraking at Venus involved a pass with the spacecraft temporarily out of contact with Earth, so the solar panels could be aligned for maximum drag.

of kinetic energy must be shed in a small time and the heating is great. For aerobraking the effect is more gradual, and *Magellan* could use its solar arrays as drag surfaces without any special coating for thermal protection. It took 70 days to reduce the apoapsis of *Magellan*'s orbit from an altitude of more than 8,000 down to 540 kilometres, a manoeuvre that would have required ten times as much chemical propellant as *Magellan* actually carried if it had been done with rocket burns. The attitude control system was used to make small tweaks to the orbit between passages through the atmosphere to avoid overheating; in the event, the surface temperatures measured on the solar panels did not exceed a modest 85 centigrade.

Surveying the surface: mappers, aeroplanes and submarines

The plan by the Russians to deploy a glider on Venus is the precursor for the more advanced idea of a powered aircraft that might travel around below the clouds making a survey of large swathes of the surface (Plate 24). Photography, of course, would be part of this, but if suitably instrumented the robot plane could also map the distribution of common minerals on the surface. *Venus Express* and *Cassini* have made a start on this kind of study. The five spectral windows in the infrared just long of visible wavelengths, shown by the VIMS instrument to be sensitive to surface spectral properties,[1] provide a potentially effective means for remotely mapping the mineralogical composition of the surface of Venus.

Figure 18.2 suggests that nightside spectral imaging of the surface from orbit in the spectral windows could be used to make preliminary global maps of olivines, pyroxenes and other ferrous materials in surface basalts. Spectra from low-flying aircraft should

---

[1] VIMS stands for the Visible and Infrared Mapping Spectrometer on the *Cassini* spacecraft, which flew past Venus in 1999, see Chapter 7.

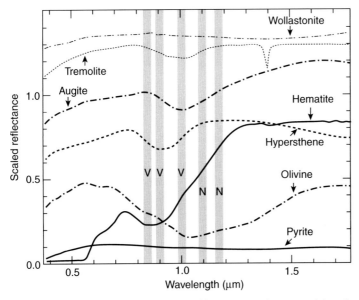

Figure 18.2 Laboratory spectra of possible Venus surface materials at low resolution, showing how they might be identified and distinguished over large areas by a near-infrared mapping instrument like VIMS. The grey bars show the atmospheric windows in which observations can be made from orbit; with an instrument on a low-flying aircraft the whole range could be covered for better discrimination between different minerals.

resolve less common materials such as haematite (iron oxide), which has already been tentatively identified in *Veneras 9* and *10* spectral reflectance data obtained at the surface, and the iron sulphides pyrite and its rarer variant pyrrhotite.

The spatial distribution of these minerals is important in order to understand what is being produced by volcanoes on Venus and the nature of the oxidation processes that change one species into another (pyrite into haematite, for example). This will be essential to quantify the rate at which volcanism releases sulphur into the atmosphere, and to understand more about how that rate may have varied over geological periods of time.

Other minerals, such as the hydrated silicate tremolite, are expected to have been produced in the wet environment, if there was one, in Venus's early history, and some of this may have survived to the present without decomposing or being covered over. Figure 18.2 shows the spectrum of tremolite and that of wollastonite (calcium silicate), the mineral thought to be regulating the amount of CO_2 in the atmosphere and hence the surface pressure. If all of these can be distinguished and mapped over the planet, it would contribute significantly to our understanding of the evolution of Venus's surface and climate and the processes involved.

Apart from 'hardening' the devices against the environment, and of course developing and deploying a low-altitude platform, suitable instrumentation for this sort of investigation already exists. However, hardening is particularly non-trivial for an instrument looking for subtle differences in infrared flux in ragingly hot surroundings. Even on Earth (and on *Venus Express*, *Galileo* and *Cassini*) such devices are cooled to reduce the background radiance and to increase the sensitivity of the detectors. On Venus it becomes a daunting task, although not impossible in principle, for refrigerators with a limited power budget when it is so hot outside.

Direct sampling of the surface

Remote measurements are great because they cover the planet, but are not an end in themselves because they can give only an approximate, general picture of the surface composition. For a better understanding of what composes the regolith, and the more pristine stuff underneath that is less weathered by exposure to the atmosphere, it is necessary to land and sample the rock and soil directly. To do this properly requires more than the hour or so that was achieved by *Venera* using insulation alone. Looking to the future, we also want some mobility to collect samples of a selection of the most interesting items at each landing site that may not be within immediate reach. It would also be good to have a long time baseline for seismometers to listen for tremors and quakes, both diagnostic of the structure of the deep interior of the planet.

There are two ways to do this in the pressure cooker environment on Venus: either develop mechanisms, motors and control electronics that can survive under high pressure and searingly hot conditions, or invent refrigerators that can keep everything cool while they work in Earthlike temperatures. For the first instruments in space, in Earth orbit, the problem was that the ambient temperatures were too low. In that case, the latter solution (thermal control) was the answer. On Venus, refrigeration is a challenge, but not impossible. At present, the most likely path seems to be that mechanisms might be made heat-resistant while electronics and detectors are cooled.

Research has been going on for some time in America and Russia towards the goal of placing instruments on Venus that can operate for long periods. A 'switched-reluctance' motor has been developed at NASA's Glenn Research Center, and under test it ran for more than a day at 8,000 revs per minute in an oven at a temperature of over 500 centigrade. High-temperature piezoelectric switches and actuators have been built that are good at temperatures over 700 centigrade. Metal semiconductors can be built that have large energy gaps tuned to Venus temperatures; field-effect transistors have been built that operated with good stability at 500 centigrade for more than 500 hours. The developers think that simple electronics for Venus are not too far off; the complex integrated circuits needed for computers and microprocessors are much more difficult. They have suggested that the latter – the 'brains' of the experiment – might be suspended on a balloon above the clouds, linked by a kind of WiFi with the mechanical devices on the surface.

NASA has also described a refrigerator, working on the well-known Stirling cycle principle, that would dump heat at Venusian rather than room temperatures. These devices are regularly used on satellites to cool to minus 200 centigrade (the temperature of liquid nitrogen, needed for most types of infrared detector used in space). There is some way to go before that will be possible on Venus, but cooling electronics to a reasonable operating temperature (say 50 centigrade) might be possible, and early tests have been encouraging if still well short of what is likely to be required.

Intrepid robot devices on Venus will also need a power source. Some sort of battery is possible in principle, perhaps topped up by the feeble solar flux that filters through the clouds, but much more promising performance is offered by a nuclear source. The radioisotope thermal generators on outer planet missions such as *Cassini*, fuelled by plutonium dioxide, run at high temperatures anyway (nearly 1,000 degrees) and could, in principle, be used on Venus.

Instruments

Once we have access to selected samples, they need to be analysed to give detailed and precise information about the geochemistry of the site. 'Detailed' means that they should be able to detect substances that are present in very small quantities, but which are nevertheless full of information about the history of the surface. This would include elements like chlorine, boron and cobalt. But of course it is necessary to know not only elemental abundances, but what chemical combination they are in. There are scores of different minerals in granite alone, each with a story to tell.

Improved versions of the X-ray fluorescence and gamma-ray spectrometers used already for composition measurements on Venus are fairly straightforward to develop, and indeed a lot of work is going on. However, much of it is focused on making the corresponding measurements on Mars, where rovers, drills and soon sample return are understandably higher up the agenda than for Venus. Mars geologists also have access to laser-induced breakdown spectroscopy, in which rocks are vaporised with a laser and the product analysed. This works best in a vacuum, or at least a thin atmosphere like that of Mars; on Venus it might require a very close approach to the rock to be sampled, a distinct disadvantage.

Rovers and drilling

Assuming we have managed to put something on Venus that can survive for weeks or months, and make the desired measurements, the next requirement is that it should be able to move around over considerable distances and not just study the samples available at a single site. This is almost bound to be cheaper and more effective than a large number of stationary landers put down at random, given the vast land area to be explored and the variety of the terrain.

There are two basic approaches that we might consider: a rover, on wheels, like those currently active on Mars, or a 'hopper' that can analyse one site, take off and fly to another to work there. Quite large hops, measured in tens of kilometres, are desirable. At each site, we will not be satisfied for long with scraping the surface, which will have been heavily modified by interactions with the corrosive atmosphere. Some kind of drill will be essential.

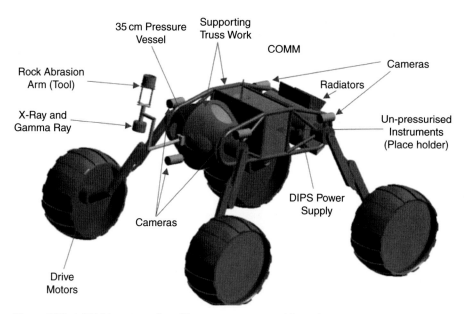

Figure 18.3 A NASA concept for a Venus rover, powered by a dynamic isotope power supply (DIPS). This is an advanced version of the radioisotope generators used on earlier planetary spacecraft, using plutonium to heat a small mechanical generator to provide electrical power to the rover.

Figure 18.3 shows a Venus rover design from NASA. About 1 metre high and 2 metres long, it has long legs that anticipate the need to trek across a rough surface. Other than that, and of course the need for high-temperature subsystems as just discussed, the general concept is familiar from the exploration of Mars.

The same cannot be said of the hopper shown in Figure 18.4. The *Venus Mobile Explorer*, as its champions at NASA's Goddard Space Fight Center call it, uses metal bellows to make a kind of robust balloon, extending when filled to provide lift to a height of around 5 kilometres, and collapsing, like a gasometer, to return to the surface at a new exploration site. Between destinations, it is carried by the wind, at typically 1 metre per second, reaching a new site a few hours later. Navigation from the Earth would be difficult, but it might be pre-programmed using a good map of where the interesting targets lie, or the robot might survey the landscape below with its cameras and decide for itself where to land after processing the information with its remote brain above the clouds.

The *Explorer* has been studied in considerable detail, to the point where it could be proposed for flight in 2023. Only two locations would be attempted the first time around, and the mission would last only five hours. The gondola, weighing nearly a ton, needs 14 cubic metres of helium at a pressure of half an atmosphere above ambient to have sufficient lift. The helium is stored at high pressure in a titanium tank, a considerable mass which is left behind as the gondola leaves the first site for the second, which will be around 10 kilometres away. During the three-hour flight, images are obtained from aloft every 1 or 2 kilometres of the traverse.

Figure 18.4 The *Venus Mobile Explorer* concept (NASA). At the high pressures and temperatures at the surface of Venus, a metallic bellows acts as a balloon that can be inflated or deflated many times to allow short hops across the surface.

Sample return: bringing material back to Earth

Venus sample return missions have been studied occasionally for the past 40 years, but all they were able to conclude was that technology developments, mainly in high-temperature systems, would be needed before they become serious contenders for flight. Only recently has the state of the art advanced to the point where this kind of mission has begun to seem realistic. At the same time, however, cost constraints have become more important. NASA has recently been studying a set of Venus sample return missions which span a range of scientific goals and which have a corresponding range of costs.

The simplest form of sample return from Venus is to fly through the upper atmosphere and return some of the gas to Earth. Studies have shown that this could be possible using a free-return ballistic trajectory. The gas sample would be captured at hypersonic velocities and would be greatly affected by the process, but we could still determine trace gas composition to very high precision back in the lab and should be able to determine elemental and isotopic ratios.

The next level of sample return would be another atmospheric mission, but would involve much more benign sampling conditions. On arrival at Venus the spacecraft would be aerocaptured into orbit and aerobraked until the orbit was circular. A sample-gathering vehicle would enter the atmosphere and slow down significantly, allowing capture of a well-mixed atmosphere sample which preserves the chemical composition. An on-board rocket would then return the sample to Venus orbit, where the orbiting spacecraft would retrieve it and bring it back to Earth. This mission is somewhat beyond NASA's low-cost *Discovery* class, though not outrageously so.

The most challenging level of sample return mission would be a selection of rocks from the Venusian surface, or even better (and harder) a core drilled from the surface and subsurface. Obviously, this will be very much more difficult and costly than an atmosphere-sampling mission, but it will happen one day because only a direct examination of Venus material will answer the key questions about the formation and evolution of the planet and the Solar System.

In 1998 the author was part of a European Space Agency (ESA) team that studied a possible Venus sample return mission. The instigator was the French Science Minister, who at the time was Claude Allègre, a geologist by profession who had a genuine interest in looking at whether Venus surface samples could be obtained in the reasonably near future at an affordable price. France has a big steer on what ESA does, so his enthusiasm gave the study extra significance over and above the usual long-term technology definition role.

We knew of course that it was a tall order to bring back scientifically useful samples from the surface using existing technology. From the first report of the ESA engineers and their industrial contacts it was clear that we would have to accept that there would be no mobility on the surface, not so much because of the wheels and motors and so on that would be required, although this would of course add weight and complexity, but rather because of the fact that just surviving the very high temperature and pressure environment for long enough to seek the best available location for sampling was beyond what we could reasonably achieve.

Providing a suitable pressure vessel for surface operations is mainly a question of addressing the mass implications, and winds should not be troublesome at about a metre per second, despite the high density. Temperature is the big problem, especially the effect on electronic subsystems inside the spacecraft. As we have seen, high-temperature electronics and refrigerators have been under development for some time, but on the more-or-less immediate timescale that was specified for the study, any mission to the surface of Venus would be restricted to the use of semiconductors with a maximum tolerance of about 40 degrees centigrade. Any kind of active refrigeration was quickly ruled out as being similarly impractical at present and the only reasonable solution is temporary protection from the environment using insulating and heat-absorbing materials.

The part of the mission carried out on the surface would have to be brief, therefore. However, a 'grab' sample from a random location would be good for a first look, especially if some rudimentary drilling or digging tool was provided to try to get material that was not weathered too much by exposure to the atmosphere. It would have to work quickly because all of the landed equipment would survive for only one hour before it overheated. Before the electronics died, the sample would have to be on its way back to Earth.

Landing and taking off on Venus present special problems. Descent all the way to the surface by parachute would require a hierarchy of chutes of decreasing sizes to stop the descent taking too long (while the electronics got hotter and hotter). On the other hand, it must not strike the surface too fast either. An ingenious solution is to inflate a balloon on the way down, regulating its size by adding a controlled amount of gas to provide the appropriate braking effect at each stage of the descent. The landing could then be relatively gentle, with an impact velocity of approximately 8 metres per second, avoiding the need for the bulky shock-absorbing system used by the *Venera* probes and replacing it with just a 10-centimetre thickness of lightweight crushable honeycomb material.

On the surface, the balloon stays attached and inflated while the sample is acquired and stored in the ascent module. This is then separated from the main body of the lander and the balloon is used again, this time to lift the sample capsule up to the cloud tops. Once at Earthlike pressures, a solid-state rocket motor would take over and propel the sample up into orbit. This is efficient in terms of limiting the amount of rocket fuel that has to be carried down to the surface, but there is a more fundamental and less-obvious reason for this approach as well. This is that any rocket, even one much larger than any that could sensibly have been used, would not have been able to reach orbit if launched from the surface of Venus. The air resistance at such high densities is much too great.[2]

The ESA study devoted a lot of effort to defining in detail how the lifetime on the surface would be maximised. The electronics and other heat-sensitive parts would be surrounded by a cocoon of high-temperature multi-insulation, manufactured by stacking and sewing together crinkled reflective foils separated by ceramic fabric. In addition, the helium from the tanks used to fill the balloon prior to take-off would flow through a heat exchanger in the cocoon to help keep it cool, and to compensate for the influx of a relatively small amount of hot CO_2 atmosphere into parts of the lander that were not kept evacuated. Finally, the electronics were to be protected by a layer of phase change material that would absorb the heat dissipated when the components are in operation as latent heat instead of a temperature rise. Lithium nitrate trihydrate is used for its high melting temperature and latent heat, and low density and volume change.

The final design for the thermal protection system included 9 kilograms of this material, and overall met the requirement of a theoretical maximum internal temperature of 65 degrees after five hours' exposition to an environment with constant temperature of 450 degrees centigrade. (This was chosen to give a comfortable margin over the primary requirement of 40 centigrade after one hour.)

Most of the time on the surface is spent drilling, to try to get below the exposed and contaminated regolith. A core sample about 2 centimetres across from a depth of at least 20 centimetres was specified, with the drill long enough to go down to 50 centimetres if the soil turned out to be soft. Clearly, even 20 centimetres would be difficult if the lander touched down on one of the flat plates of hard rock that are common on Venus, but in that case a centimetre or two would be enough as the weathered crust would be correspondingly thin.

A vacuum cleaner-like arrangement would sweep up and collect some nearby dust and gravel to provide a back-up sample, one that is definitely weathered and so complementary to the more pristine material from the drill. In a manner that was not specified at this time (the study was short – just four months – and left a lot of detail for later work if the mission progressed further in the Agency's plans), the samples would be transferred to the

---

[2] An analogous problem is the launch of Polaris-type missiles from submerged submarines on Earth. If a film of one of these deploying is studied carefully, you can see that the missile is pushed from its storage tube by a blast of compressed air, and the rocket motor is not fired until it breaks through the sea surface.

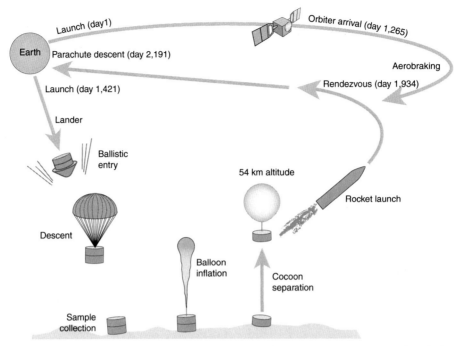

Figure 18.5 An overview of the European Venus sample return mission concept. The whole sequence takes about six years, and requires two of the largest *Ariane V* launcher vehicles.

ascent capsule. The capsule and its rocket motor detach from the lander with its drill and are pulled aloft by the balloon, now fully inflated (Figure 18.5).

The balloon obviously has to be tough. Suitable high temperature and acid resistant material is available, but the size of the envelope was such that mass became a key issue. When considering density, mechanical properties at high temperature and the possibility to manufacture thin tight films, the most promising candidate turned out to be a material called polyphenylene-benzobis-oxazole. This is commercially available, and actually offers a safety factor of about ten for temperatures up to 650 centigrade, well in excess of what is required. Of course, before it could be used a complete characterisation in a simulated Venus environment would have to be carried out, including inflation after being tightly packed and stored for perhaps a year on Earth, and for several months in space on the cruise to Venus.

Once the ascent package has travelled through the upper boundary of the clouds, the pressure is less than one atmosphere and the first solid-fuel rocket fires. Three stages are used to get to orbit, an arrangement that optimises the use of fuel and also allows the use of only the relatively light third stage to navigate into a suitable orbit around Venus. The fuel obviously has to be something special to survive the trip to the surface and back without exploding, and still deliver enough thrust to do the job. Rather incredibly, such stuff does exist; the study engineers came up with a mixture of hydrazinium nitroformate and glycidyl azide polymer with aluminium powder.

Delivering the whole thing from Earth to Venus would, by now, be fairly routine, although the payload is big and heavy and would require ESA's largest rocket, the *Ariane V*. Getting the sample back is quite another matter. How could this be done? The 'rockoon' arrangement already described could get the sample to Venus orbit, but there is no way it would have the fuel, or the navigation equipment, to get back to Earth. The answer would have to be an *Apollo*-style rendezvous in orbit with a mother ship. The trouble is, this does not add up in terms of mass, and the team ended up recommending *two* launches, one *Ariane V* to deliver the lander, sampling and ascent vehicle, and another to deliver the return vehicle to orbit where it would wait for the sample to arrive, collect it and then fly it home.

An obvious problem with all of this is how the two spacecraft would find each other. The answer (it turns out) is slowly; matching orbits so that the two coincide can be done with a fairly low cost in fuel, provided you are prepared to spend a year or two doing it. This would not be viable if the craft were manned, of course; then, big burns with powerful engines are the only way. But inanimate samples can take their time. The net result is that the whole mission takes six years from launch until arrival of the sample of Venus soil back on Earth. The total expected cost was not divulged by ESA (because a more detailed study would be needed to refine the numbers) but a good guess would be in the region of €4 billion.

Chapter 19
Beyond the horizon: human expeditions

When, if ever

When plans for exploring the Solar System with manned spacecraft are discussed, Venus tends to get short shrift. In the near term, of course it is natural to talk about a return to the Moon and the establishment of a manned base there. The short journey times and low gravity are just two of many reasons this is the easiest and least expensive option for human exploration in the near term. It is also a good place to practise survival techniques and develop procedures for living successfully in space before venturing further afield. Also, of course, there is much about the Moon that is of scientific and practical interest that makes exploration still a valid objective 40 years after the first *Apollo* landing.

Once humans are permanently established on the Moon, almost everyone thinks of pressing on to Mars. Often the two are programmatically linked, with the lunar landings seen explicitly as a stepping stone on the way to the red planet. This was the case for the NASA initiative started under President George W. Bush, and recently terminated by President Obama. Mars remains a long-term goal in Europe under the *Aurora* programme. The reasons for favouring Mars over Venus are pretty much taken for granted: men and women can land there and explore in the traditional manner, driving buggies, using hammers and drills, climbing mountains and cliffs or descending into deep valleys. It is fairly easy to see, in outline at least, how they could build permanent bases and live in them, becoming self-supporting by growing food and mining ice deposits for water, possibly even making their own rocket fuel. No one would think of trying any of that on Venus.

It is possible to imagine, as writers of science fiction have done, a manned base on the surface of Venus in which the intrepid pioneers occupy buildings cooled to a comfortable 20 degrees centigrade by an advanced refrigeration system that dumps heat into the surrounding environment at 450 centigrade. They might even have suits and vehicles that allow them the mobility to explore with the same kind of protection against the heat and the pressure. This equipment need not violate the laws of physics, but it is so far in advance of our current technologies that we can only dream about it at present.

If we rule out serious thinking about humans living on the surface of Venus, what is left to look forward to after robots have done the first few decades of serious exploration and men and women have walked on Mars? Obviously, there is no particular technical problem with sending a crew to fly past Venus or go into orbit and conduct observations from there, and in fact NASA did a detailed study of such a mission nearly 50 years ago, in the days of the *Apollo* lunar landings. The reason it never happened, and will not for as far as we can see into a realistic future space programme, is the high cost relative to using robots. Many scientists would argue that robots do a better job anyway. Nevertheless, the urge to explore in person will lead future generations to pick up this particular thread in something like 50 years' time.

After that, we could easily imagine a manned platform floating near the top of the cloud deck. As we have seen, the temperatures and pressures there are Earthlike and so the technological challenges are not beyond our current powers. The project could take place quite soon given the will, and rather a lot of money. It seems likely that a Venus-orbiting manned space station would be a logical precursor.

After that, it gets very difficult to break new ground. Eventually we will wish to do literally that, but exactly how we can only dimly perceive. One possibility, much discussed, is that instead of trying to survive the conditions on the surface of Venus, we might change them to something more Earthlike. This process, usually called terraforming, would obviously require massive feats of geoengineering if it is possible at all. Just thinking about it is so much fun that whole conferences have been devoted to discussing concepts and arguing about how best it might be attempted in some distant future. The appeal, of course, is that a terraformed Venus could be an attractive venue for human migrants to actually live and thrive in large numbers, something that cannot really be said for Mars or indeed anywhere else in the Solar System.[1] So, as we come to the final chapter of our review of Venus science and peer into the future, trying to keep our feet on the ground as we do so, a brief discussion of the state of the terraforming debate is worth including.

NASA's 1967 manned flyby

The study was carried out for the Manned Space Flight division of NASA by Bellcomm Inc.[2] The basic idea was to use the large *Saturn V* launcher to send a crew of three in a modified *Apollo* capsule to fly around Venus and return to Earth. The fact that there was no landing vehicle, as there was for the lunar mission, allowed additional living quarters to be attached to the control module for the comfort of the astronauts on the much longer voyage. Figure 19.1 shows one of the surviving blueprints for the spacecraft, and although much of the detail is not legible when reduced to this scale, it does give the impression everything was ready to build and fly.

---

[1] Many enthusiastic Mars terraformers will disagree with this, but even allowing for the speculative nature of the argument, a smaller planet further from the Sun will be much harder to keep Earthlike than one the same size and warmer.

[2] Bellcomm was a subsidiary of the telecommunications giant AT&T that was set up in 1963, primarily to support NASA through contracts for advanced technical studies like this Venus mission.

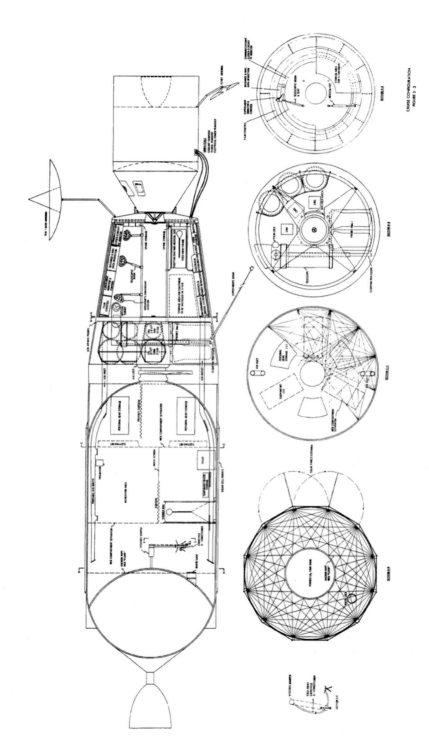

Figure 19.1 The detail in this technical drawing of the Venus flyby spacecraft envisioned by Bellcomm in 1967 gives a sense of the advanced state of the project when it was cancelled.

Launch was anticipated for 1973, with an outward flight time of 123 days. During this time, and the longer return leg of 273 days, the crew were to spend ten hours of each day making scientific observations with the suite of instruments carried on board. With an estimated mass of 1,500 kilograms, these included a 200-kilogram probe to be dropped into the atmosphere and a large optical telescope with a 40-centimetre aperture. The latter would take pictures (on film, of course; 50 pounds of payload mass was set aside for this, enough for an estimated 10,000 photographs) and would also have an infrared and ultra-violet capability. If it had flown, the probe would probably have failed before reaching the surface, because the design study assumed that the density there was less than a tenth of its true value.

Given the early date of the study, it is interesting to see what they saw as the most important goals. Here are the key objectives as presented in the final report, dated 1967:

- Atmospheric density, temperature and pressure as functions of altitude, latitude and time.
- Definition of the planetary surface and its properties.
- Chemical composition of the low atmosphere and the planetary surface.
- Ionospheric data such as radio reflectivity and electron density and properties of cloud layers.
- Optical astronomy – ultraviolet and infrared measurements above the Earth's atmosphere to aid in the determination of the spatial distribution of hydrogen.
- Solar astronomy – ultraviolet, X-ray and infrared measurements of the solar spectrum and space monitoring of solar events.
- Radio and radar astronomy – radio observations to map the brightness of the radio sky and to investigate solar, stellar and planetary radio emissions; radar measurements of the surface of Venus and Mercury.
- X-ray astronomy – measurements to identify new X-ray sources in the galactic system and to obtain additional information on sources previously identified.
- Data on the Earth–Venus interplanetary environment, including particulate radiation, magnetic fields and meteoroids.
- Data on the planet Mercury, which will be in mutual planetary alignment with Venus approximately two weeks after the Venus flyby.

With the probe data stored on magnetic tape, the spacecraft would swing around Venus on a course that redirected it back to Earth. On arrival, the command module would be separated from the living quarters and used to return the crew through the Earth's atmosphere for an *Apollo*-style splashdown in the Pacific Ocean. It would then be 1 December 1974.[3]

Floating research stations

In 2074 (say) we might see manned scientific outposts on Venus in the form of buoyant stations that float just above the clouds, where temperatures and pressures, and gravity,

---

[3] The actual headline in the newspapers on that day was not the return of astronauts from Venus, but the crash of TWA Flight 514, a Boeing 727, at Dulles Airport near Washington DC which killed all 92 people on board.

very close to those on Earth's surface are to be found. Thin walls would suffice therefore, and a breathable mixture of oxygen and nitrogen inside would provide buoyancy in the CO_2 atmosphere comparable to that of helium balloons on Earth (Plate 25). The intense sunlight would provide power, and the atmospheric circulation would provide a day–night cycle about four times as long as Earth's (a substantial improvement over the 243-day solar cycle at the surface).

Robots would be dispatched to survey and sample the lower atmosphere and terrain, and the floating stations could provide a convenient staging post for laboratory analysis and sample return to Earth. It all sounds quite marvellous and relatively free of technical show-stoppers given some plausible advances in electronics, materials and propulsion in the next few decades. The problem again is likely to be willpower, since the costs and risks are undeniably substantial.

Terraforming Venus: building an Earthlike climate

Skipping ahead to perhaps 2274, giving short shrift in passing to any idea that there might be a human colony on the surface of Venus while it still has its present inhospitable climate, let us look at ways in which our descendants might have altered the environment to suit themselves better. Obviously we must assume that technology will continue to make massive advances and that humans will work together to make truly world-scale efforts to improve the lot of mankind. Geoengineering to stabilise Earth's climate will probably be a reality by this time; so probably will such a shortage of real estate on the home planet that a second, nearby, habitable world will seem not only a romantic or scientific goal but a practical necessity.

What, then, will they do? The idea of ever terraforming Venus was first advanced in a serious way, and given respectability as a legitimate topic for discussion, by no less a scientist than Carl Sagan. In a paper in 1961, he proposed seeding the clouds on Venus with synthetic bacteria or algae that would reproduce and gradually consume the carbon dioxide via photosynthesis, converting it to oxygen and solid organic matter. Later, as the true nature of Venus's atmosphere was progressively revealed (thanks in no small part to the efforts of Sagan himself, as we have seen), he realised that this approach was unlikely to work, even in theory. Among the reasons are the shortage of water to provide the hydrogen needed to make organic molecules, the acid environment in the clouds, and the hot surface conditions that would decompose the organics and release the carbon dioxide again. Although it keeps cropping up in popular discussions and journalism, the whole idea of terraforming Venus by biological agents currently lacks any plausible scientific basis.

Another idea, one that appears regularly in futuristic views of how we might combat global warming on the Earth, is to shield the planet from the Sun to the point where it would cool down substantially. Balloons, chaff or thin films of highly reflective material stretched on lightweight frames might be placed in orbit and used to increase the effective albedo of the planet. While the (hopefully) relatively small tweaks to the insolation needed to protect the Earth are no doubt possible, although there could be unintended consequences that are not well understood, to make a sufficient difference on Venus would be a massive challenge.

One much-discussed idea that has considerable merit is to place the reflector near the Lagrangian point between Venus and the Sun,[4] where the attenuation achieved could be constant and carefully controlled. However, when the calculations are done it turns out that the screen would have to be larger across than Venus itself. Even this is not inconceivable, as materials of great thinness and strength are becoming available, and in space the support structure required is minimal. For example, a circular shield could be kept rigid by spinning it slowly, without any kind of frame at all.

Assuming the engineering problems could be solved, most of the studies go on to assume that Venus would cool until it became Earthlike. This is not necessarily true. The predicted surface temperature as a function of the attenuation provided by the shield, assuming that any change takes place so slowly that the system comes to equilibrium, decreases in a series of steps. The first major shift in the gradual cooling expected by the model is the loss of the cloud cover, which allows more of the remaining sunlight to reach the surface. Eventually the freezing point of carbon dioxide is reached and the atmosphere collapses, stopping at a surface pressure of around 3 atmospheres of nearly pure nitrogen on top of a surface crust of frozen CO_2 several kilometres thick. If the reduction in insolation continues, the nitrogen also freezes and the exercise becomes meaningless; the surface of Venus now resembles that of Triton and is just about as benign for human habitation. Even the frozen CO_2, gaseous N_2 state is not much use to us as a habitat.

The transition zone between gaseous CO_2 (too hot) and solid CO_2 (too cold) is not one than we are likely to be able to manipulate to the desired 20 centigrade or so. An added complication is the release of interior heat; volcanoes would cause the CO_2 ice to sublime back into gas and some of the CO_2 might survive in the atmosphere long enough to raise the mean temperature again. Modelling and regulating this requires much more knowledge about the interior of Venus and the volcanic regime at the surface than we possess at present, but we can say with some assurance that Venus, if 'terraformed' with sunshields alone, is unlikely to be the colonists' paradise we would like to see.

It would be better to remove the CO_2 from the atmosphere by other means, either chemical or physical. If a future super-civilisation could condense the CO_2 and ship it out from the planet *en masse*, then obviously an Earthlike atmosphere could be constructed and maintained by artificial means. If this exactly resembled the Earth, the mean surface temperature would be about 45 centigrade at Venus's distance from the Sun, and there would be a comfortably habitable zone at high latitudes (and no seasons).

A more practical, although still formidable, approach would be to deliver large amounts of water to facilitate the conversion of atmosphere carbon dioxide into carbonate rock, the same process that removed most of the Earth's early inventory of atmospheric CO_2. Imaginative but potentially realistic proposals for achieving this have included diverting a number of large comets into collisions with Venus, or even towing one of the icy satellites of the outer planets to Venus. If we become capable of cosmic billiards on this

---

[4] The Lagrangian point is where the gravitational attraction due to Venus is the same as that due to the Sun. A shield placed there could maintain its position indefinitely and provide permanent shade.

kind of scale, we might also be able to arrange the collision in such a way that the spin of Venus on its axis is increased to something nearer the familiar 24 hours. It could also blast away some of the carbon dioxide and nitrogen, to give us a flying start. Even assuming we could do all of that, and there are all sorts of problems over and above the sheer scale of the operation, we would have to wait many hundreds of millions of years before moving in to our new home.

Epilogue

Setting aside the Sun and the Moon, Venus is the brightest and the prettiest object in the sky. It is the closest planet, and the most Earthlike in size and structure. It has a thick atmosphere, and a serious case of greenhouse warming that resembles in many details the process threatening to cause serious global change on Earth. The surface has mountains, rivers and seabeds, but no water. The atmosphere contains familiar gases, nitrogen, carbon dioxide, water vapour.

As humankind gets more and more technically proficient, we explore Venus in greater detail and we learn much about our origins and the history of the Solar System that we inhabit. We will fly to Venus in person eventually, first to fly around the planet and return, then to orbit, and then to float in the Earthlike conditions near the cloud tops. We may even descend to the surface, in bathysphere-type craft with some advanced system of thermal control and a strong hull. Finally, we may become so godlike that we restore Venus to its earlier, more Earthlike state and move in there. There is nowhere else that seriously offers that possibility, not even Mars, until we reach the stars and their planets.

In the meantime, we should take some satisfaction from living in the time when other worlds like this have finally been revealed to us, so that at last we understand them. This has been the story of that understanding and how we acquired it, a story that is still not yet and never will be complete.

References and acknowledgements

Listed here are the sources for figures and other material used in the book, in sequence from the first chapter, and some guidance for further reading about Venus. Figures not listed here either originated with one of the space agencies, NASA, Roscosmos, ESA or JAXA, as will be obvious from the context, or were produced for this book by Dr D.J. Taylor.

Chapter 1

The beautiful photographs of Venus in Figure 1.2 were obtained in 2002 by Chris Proctor at the Torquay Boys' Grammar School Observatory in Devon, England.

Excellent detailed discussions of telescopic observations of Venus, from the earliest times until recently, are to be found in

Moore, Patrick 2002. *Venus*, London: Cassell & Co.

Patrick's sketch of the ashen light in Figure 1.6 appears in this book.

The extracts from the notes by Jeremiah Horrocks are quoted in *Memoir of the Life and Labours of the Reverend Jeremiah Horrox* (sic.), Whatton, A.B., 1859.

For the full calculation of Halley's method for determining the Earth–Sun distance, see http://www-istp.gsfc.nasa.gov/stargaze/Svenus1.htm (accessed 24 January 2014). The quote attributed to Mikhail Lomonosov, describing the first published observation of an atmosphere on Venus, dates from 1761 and was then of course in Russian. This English translation appears in

Maunder, M. and Moore, P. 'Transit: When Planets Cross the Sun', p.38, London, 1999.

William Herschel's remarks on the nature of Venus's atmosphere are from a paper read to the Royal Society of London on June 13, 1793.

Percival Lowell's sketches in Figure 1.5 and the accompanying remarks are from

Lowell, P. 1897, 'Rotation period and surface character of Venus', *Monthly Notices of the Royal Astronomical Society*, vol. 57, 148–149,

and the quote accompanying Figure 1.5 is from

Lowell, P. 1897 'The Rotation Period of Venus', *Astromische Nachrichten, 3406, 24–25.*

The sketches of the Venusian ultraviolet markings in Figure 1.7 are from

Boyer, C. and Guérin, P. 1969. 'Etude de la rotation rétrograde, en 4 jours, de la couche extérieure nuageuse de Vénus', *Icarus* 11: 338–355.

The water vapour measurements in Figure 1.8 are presented and discussed in detail in

Barker, E. 1975. 'Observations of Venus water vapor over the disc of Venus: the 1972–74 data using the H2O lines at 8197 and 8176 Å', *Icarus* 25: 268–281.

Chapter 2

A detailed account of the early days of the *Mariner* programme was written by:

Koppes, Clayton R. 1982. '*JPL and the American Space Program: A History of the Jet Propulsion Laboratory*', New Haven, CT: Yale University Press.

The calculations in Figure 2.1 are from an internal report by members of the *Mariner 2* project team at JPL. For some history on the *Venera* programme, see www. Russianspaceweb.com/Venera

Chapter 3

A comprehensive account of the *Pioneer Venus* mission and its results can be found in the special issue:

Journal of Geophysical Research: Space Physics 85/A13: 30 December 1980.

Also in

Hunten, D., Colin, L., Donahue, T. and Moroz, V. (eds.) 1982. *Venus*, Tucson: University of Arizona Press.

Chapter 4

The early radar images of Venus obtained with large radio dishes on the Earth are presented in

Goldstein, R. 1967. 'Radar Studies of Venus', in *Moon and Planets*, A. Dollfus ed., Amsterdam: North-Holland.

For a detailed account of the *Magellan* Venus radar mapping mission and its results, see the special issue:

Journal of Geophysical Research: Solid Earth 95/B6: 10 June 1990.

Figure 4.3, showing the height profiles of Earth, Mars and Venus, is from

Lang, Kenneth R. 2003. *The Cambridge Guide to the Solar System*. Cambridge University Press, Cambridge.

Chapter 5

The various mission studies and proposals planned for Venus are mostly written up only in the grey literature, but good summaries can be found on the website for the Venus Exploration Analysis Group:
http://www.lpi.usra.edu/vexag/ (accessed 24 January 2014).
The European Space Agency planning document referenced in the text is

ESA, 1994. Horizon 2000+: European Space Science in the 21st Century. ESA SP-1180, Noordwijk.

Chapter 6

The discovery of the near infrared spectral 'windows' was reported in

Allen, D. A. and Crawford, J. W. 1984. 'Cloud structure on the dark side of Venus', *Nature* 307: 222–224.

and the theoretical spectra in Figure 6.2 are from

Kamp, L. W., Taylor, F. W. and Calcutt, S. B. 1988, 'Structure of Venus's atmosphere from modelling of night side infrared spectra', *Nature* 336: 360–362.

and further discussed in

Kamp, L. W. and Taylor, F. W., 1990. 'Radiative transfer models of the night side of Venus', *Icarus* 86: 510–529.

The ground-based spectrum of the night side of Venus in the 2.3 micron spectral 'window' shown in Figure 6.5 was obtained at the Infrared Telescope in Hawaii in 1994, reported in

Bézard, B., de Bergh, C., Crisp, D. and Millard, J.-P. 1990. 'The deep atmosphere of Venus revealed by high resolution nightside spectra', *Nature* 345: 508–511.

Chapter 7

The *Galileo* Near Infrared Mapping Spectrometer (NIMS) surface observations were reported by

Carlson, R. W., Baines, K. H., Girard, M., Kamp, L. W., Drossart, P., Encrenaz T. and Taylor, F. W. 1993. 'Galileo/NIMS near-infrared thermal imagery of the surface of Venus', *Lunar and Planetary Science Conference XXIV Abstracts*: 253.

The measurements of carbon monoxide and their interpretation are in

Collard, A. D., Taylor, F. W., Calcutt, S. B., Carlson, R. W., Kamp, L., Baines, K., Encrenaz, Th., Drossart, P., Lellouch, E. and Bézard, B. 1993. 'Latitudinal distribution of carbon monoxide in the deep atmosphere of Venus', *Planetary and Space Science* 41/7: 487–494.

and

Taylor, F. W. 1995. 'Carbon monoxide in the deep atmosphere of Venus', *Advances in Space Research* 16/6: 81–88.

The presentation of *Cassini*-VIMS observations of Venus and the original versions on which Figures 7.6 and 18.2 are based are in

Baines, K. H., Bellucci, G., Bibring, J.-P., Brown, R. H., Buratti, B. J., Bussoletti, E., Capaccioni, F., Cerroni, P., Clark, R. N., Coradini, A., Cruikshank, D. P., Drossart, P., Formisano, V., Jaumann, R., Langevin, Y., Matson, D. L., McCord, T. B., Mennella, V., Nelson, R. M., Nicholson, P. D., Sicardy, B., Sotin, C., Hansen, G. B., Aiello, J. J., Amici, S., 2000. 'Detection of sub-micron radiation from the surface of Venus by Cassini/VIMS', *Icarus* 148: 307–311.

Chapter 8

For descriptions of *Venus Express* and its experiments, see

Taylor, F. W. (ed.) 2006. *Planetary and Space Science*, special issue, 'The planet Venus and the Venus Express mission', 54/13–14: 1247–1496.

For an overview of the results see

Titov, D. V. (ed.) 2009. 'Venus Express: results of the nominal mission', *Journal of Geophysical Research* 115/E5 and E9.

and

Svedhem, H., Witasse, O., Sohl, F., Titov, D. and Grinspoon, D. (eds.) 2011. *Planetary and Space Science*, special issue 'Comparative planetology: Venus-Earth-Mars', 59/10: 887–1112.

All of these are special issues of the journals indicated, with many articles by a wide range of authors.

The cartoon in Figure 8.9 showing attempts to detect lightning in the clouds is courtesy of Prof. Chris Russell.

The details, including figures, of the *Akatsuki* mission are courtesy of Takeshi Imamura and Takehiko Satoh of JAXA.

Chapter 9

For a recent overview of the surface geochemistry measurements made by the *Venera* and *VEGA* landers, see the review by Allan Trieman in

Exploring Venus as a Terrestrial Planet 2007, Geophysical Monograph No. 176, American Geophysical Union.

For details of the *Magellan* radar images and their interpretation, see the massive

Magellan special edition, January 1992. *Journal of Geophysical Research*, volume 97, part E10.

The details about noble gas isotopic ratios that appear in Figure 9.2 were compiled by Kevin Baines and the VALOR team.

Chapter 10

The topics in this chapter are covered in more depth in

Taylor, F. W. 2010. *Planetary Atmospheres*, Oxford University Press.

For a discussion of radiation in the upper atmosphere and Figure 10.8 see

Lopez-Puertas, M. and Taylor, F. W. 2002. *Non-Local Thermodynamic Equilibrium in Atmospheres*. London: World Scientific Publishing.

The upper atmospheric measurements by Pioneer Venus are presented by

Keating, G. M., Taylor, F. W., Nicholson, J. Y. and Hinson, E. W. 1979. 'Short-term cyclic variations of the Venus upper atmosphere', *Science* 205: 62–65.

and the discussion of the energy balance of Venus by

Schofield, J. T. and Taylor, F. W. 1982. 'Net global thermal emission from the Venus atmosphere', *Icarus* 52: 245–262.

Chapter 11

Several of the ideas in this chapter are from discussions with many people including Dr Richard Ghail of Imperial College, London, and the authors of the following papers:

Esposito, L. W. 1984. 'Sulfur dioxide: Episodic injection shows evidence for active Venus volcanism', *Science* 223: 1072–1074.

Hashimoto, G. L. and Abe, Y. 2005. 'Climate control on Venus: comparison of the carbonate and pyrite models', *Planet. Space Science* 53: 839–848.

Head, J., Crumpler, L. Aubele, J. Guest, J. and Saunders, R. 1992. 'Venus volcanism: classification of volcanic features and structures, associations, and global distribution from Magellan data', *Journal of Geophysical Research* 97(E8): 13153–13197.

Marcq, E., Bertaux, J-L., Montmessin, F. and Belyaev, D. 2012. 'Variations of sulphur dioxide at the cloud top of Venus/'s dynamic atmosphere', *Nature Geoscience*. doi:10.1038/ngeo1650.

Taylor, F. W. and Grinspoon, D. H. 2009. 'Climate evolution of Venus', *Journal of Geophysical Research* 114: E00B40, doi:10.1029/2008JE003316.

Zhang, X., Liang, M-C., Montmessin, F., Bertaux, J-L., Parkinson, C. and Yung, Y. L. 2010. 'Photolysis of sulphuric acid as the source of sulphur oxides in the mesosphere of Venus', *Nature Geoscience*. doi:10.1038/ngeo989.

The chapter contains quotes from

Solomon, S. and Head, J. 1982. 'Mechanisms for lithospheric heat transport on Venus: Implications for tectonic style and volcanism', *Journal of Geophysical Research* 87(B11), 1393–1396.

and from

Knollenberg, R. G. and Hunten, D. M. 1980. 'The microphysics of the clouds of Venus: Results of the Pioneer Venus nephelometer experiment', *Journal of Geophysical Research* 85(A13), 8,039–8,058.

Grinspoon, D. H., Pollack, J. B., Sitton, B. R., Carlson, R. W., Kamp, L. W., Baines, K. H., Encrenaz, T. and Taylor, F. W. 1993. 'Probing Venus's cloud structure with Galileo NIMS'', *Planetary and Space Science* 41: 515–542.

Chapter 12

A recent summary of current knowledge of the cloud structure on Venus, with emphasis on recent progress using *Venus Express*, is in the doctoral thesis by Joanna Barstow (Oxford University, 2012), summarised in the paper:

Barstow, J. K., Tsang, C. C. C., Wilson, C. F., Irwin, P. G. J., Taylor, F. W., McGouldrick, K., Drossart, P., Piccioniand, G. and Tellmann, S. 2012. 'Models of the global cloud structure on Venus derived from Venus Express observations', *Icarus* 217/2: 542–560.

A review with emphasis on the chemistry of the cloud production and loss cycle appears in

Mills, F. P., Esposito, L. W. and Yung, Y. L. 2007. *In Venus as a Terrestrial Planet*, AGU Geophysical Monograph No. 176.

For an earlier but still valuable account of many aspects of Venus cloud studies, see

Esposito, L. W., Knollenberg, R. G., Marov, M. Ya., Toon, O. B. and Turco, R. P. 1983. 'The clouds and hazes on Venus', in *Venus*, ed. D. M. Hunten, L. Colin, T. M. Donahue and V. I. Moroz, Tucson, University of Arizona Press.

and

Esposito, L. W., Bertaux, J-L., Krasnopolsky, V., Moroz, V. I. and Zasova, L. V. 1997. 'Chemistry of lower atmosphere and clouds' in *Venus* 2, ed. S. W. Bougher, D. M. Hunten and R. J. Phillips, Tucson, University of Arizona Press.

Figure 12.8 is adapted from

Carlson, R. W., Kamp, L. W., Baines, K. H., Pollack, J. B., Grinspoon, D. H., Encrenaz, T., Drossart, P., Taylor, F. W. 1993. 'Variations in Venus cloud-particle properties: a new view of Venus's cloud morphology as observed by the *Galileo* near infrared mapping spectrometer', *Planetary and Space Science* 41: 477–485.

The polarisation calculations in Figure 12.1 are described by

Hansen, J. E. and Hovenier, J. W., 1974. 'Interpretation of the polarization of Venus', *Journal of Atmospheric Sciences* 31: 1137–1160.

Chapter 13

Figure 13.9 is from the *Mariner 10* imaging team report on Venus dynamics in:

Murray, B. C., Belton, M. J., Danielson, G. E., Davies, M. E., Gault, D., Hapke, B., O'leary, B., Strom, R. G., Suomi, V. and Trask, N. 1974. 'Venus: atmospheric motion and structure from Mariner 10 pictures', *Science* 183(4131): 1307–15.

Venus Express cloud-tracked wind measurements are presented in

Sanchez-Lavega, A., Hueso, R., Piccioni, G., Drossart, P., Peralta, J., Perez-Hoyos, S., Wilson, C. F., Taylor, F. W., Baines, K. H., Luz, D., Erard, S. and Lebonnois, S. 2008. 'Variable winds on Venus mapped in three dimensions', *Geophysical Research Letters* 35/ L13204, doi:10.1029/2008GL033817.

The carbon monoxide abundance determinations in Figure 13.6 are from:

Tsang, C. C. C., Irwin, P. G. J., Taylor, F. W., Wilson, C. F., Lee, C., de Kok, R., Drossart, P., Piccioni, G., Bezard, B., Calcutt, S. and Venus Express/VIRTIS Team 2008. 'Tropospheric Carbon Monoxide Concentrations and Variability on Venus from Venus Express/VIRTIS-M Observations', *Journal of Geophysical Research* 113/E00B08, doi:10.1029/2008JE003089.

Chapter 14

The classic publications on long-term climate change on Venus are

Bullock, M. A. and Grinspoon, D. H. 1996. 'The stability of climate on Venus', *J. Geophys. Res.*, 101: 7521–7530.

Bullock, M. A. and Grinspoon, D. H. 2001. 'The recent evolution of climate on Venus', *Icarus* 150: 19–37.

This chapter is based on a more recent study, see

Taylor, F. W. and Grinspoon, D. H. 2009. 'Climate evolution of Venus', *Journal of Geophysical Research* 114, E00B40, doi:10.1029/2008JE003316.

The basic physics employed in the climate models may be found in the author's textbooks,

Planetary Atmospheres, Oxford University Press, 2010

and

Elementary Climate Physics, Oxford University Press, 2005.

Both are at a level aimed at undergraduate physics students.

Chapter 15

Unlike most of the topics in this book, this author has not spent much time contemplating Life on Venus, and should therefore particularly acknowledge the contributions of others. Much of this chapter draws on the following three papers, and related presentations and discussions with the authors and their colleagues[1]. The original papers are a good place to start for any reader who wants to explore Venus exobiology in more depth.

Cockell, C. S. 1999: Life on Venus. *Planetary and Space Science* 47, 12, 1487–1501.

Grinspoon, David H. and Bullock, Mark A. 2007. 'Astrobiology and Venus exploration' in *Exploring Venus as a Terrestrial Planet*, ed. L. W. Esposito, E. R. Stofan and T. E. Cravens, AGU Geophysical Monographs, Vol. 176.

Schulze-Makuch, D. and Irwin, L. N. 2002. 'Reassessing the possibility of life on Venus: Proposal for an Astrobiology Mission', *Astrobiology* 2/2: 197–202.

The list of 'next steps' for understanding the exobiology of Venus is based on one I wrote down at a talk by Dr David Grinspoon.

Chapter 16

Programme and mission planning discussions are often behind closed doors and, even when published in the grey literature, subject to frequent change. This chapter could be much, much longer, but is condensed into a summary that gives the flavour of recent discussions about Venus.

Chapter 17

Specific mission proposals present as a curious blend of publicity (to garner support) and secrecy (because the selection is competitive). The former is most open (and redacted) when in the form of presentations at international conferences; the actual proposals submitted to the funding bodies are usually treated as privileged information.

The microprobe design in figure 17.4 comes from a design study by QinetiQ Ltd on which the author was a consultant.

---

[1] That is, those who admit to being exobiologists, something which is becoming more common than it used to be. Coming out of that particular closet still requires courage so far as Venus is concerned however.

Chapter 18

Technology development activities are also quite opaque and volatile so this is again a very condensed summary.

Figure 18.2 is from the paper by Baines et al. 2000, referenced above (Chapter 7).

Chapter 19

The output from the 1967 design by Bellcomm for a manned mission to Venus is owned by NASA, who commissioned the study. Bellcomm itself ceased to exist in 1972.

Appendix A
Chronology of space missions to Venus

Name	Country	Launch Date	Type of Mission	Notes
Sputnik 7	USSR	4 Feb 1961	Venus Impact	Failed
Venera 1	USSR	12 Feb 1961	Venus Flyby	Contact Lost
Mariner 1	USA	22 July 1962	Venus Flyby	Launch Failure
Sputnik 19	USSR	25 Aug 1962	Venus Flyby	Failed
Mariner 2	USA	27 Aug 1962	Venus Flyby	Successful
Sputnik 20	USSR	1 Sept 1962	Venus Flyby	Failed
Sputnik 21	USSR	12 Sept 1962	Venus Flyby	Failed
Cosmos 21	USSR	11 Nov 1963	Venera Test Flight?	
Venera 1964A	USSR	19 Feb 1964	Venus Flyby	Launch Failure
Venera 1964B	USSR	1 March 1964	Venus Flyby	Launch Failure
Cosmos 27	USSR	27 March 1964	Venus Flyby	
Zond 1	USSR	2 April 1964	Venus Flyby	Contact Lost
Venera 2	USSR	12 Nov 1965	Venus Flyby	Contact Lost
Venera 3	USSR	16 Nov 1965	Venus Lander	Contact Lost
Cosmos 96	USSR	23 Nov 1965	Venus Lander?	Failed
Venera 1965A	USSR	23 Nov 1965	Venus Flyby	Launch Failure
Venera 4	USSR	12 June 1967	Atmospheric Entry Probe	Successful
Mariner 5	USA	14 June 1967	Venus Flyby	Successful
Cosmos 167	USSR	17 June 1967	Atmospheric Entry Probe	Failed
Venera 5	USSR	5 Jan 1969	Atmospheric Entry Probe	Successful
Venera 6	USSR	10 Jan 1969	Atmospheric Entry Probe	Successful
Venera 7	USSR	17 Aug 1970	Surface Lander	Successful
Cosmos 359	USSR	22 Aug 1970	Atmospheric Entry Probe	Failed
Venera 8	USSR	27 March 1972	Atmospheric Entry Probe	Successful
Cosmos 482	USSR	31 March 1972	Atmospheric Entry Probe	Failed
Mariner 10	USA	4 Nov 1973	Venus/Mercury Flybys	Successful
Venera 9	USSR	8 June 1975	Orbiter and Lander	Successful
Venera 10	USSR	14 June 1975	Orbiter and Lander	Successful
Pioneer Venus 1	USA	20 May 1978	Orbiter	Successful
Pioneer Venus 2	USA	8 Aug 1978	Multiple Entry Probes	Successful
Venera 11	USSR	9 Sept 1978	Orbiter and Lander	Successful

Name	Country	Launch Date	Type of Mission	Notes
Venera 12	USSR	14 Sept 1978	Orbiter and Lander	Successful
Venera 13	USSR	30 Oct 1981	Orbiter and Lander	Successful
Venera 14	USSR	4 Nov 1981	Orbiter and Lander	Successful
Venera 15	USSR	2 June 1983	Orbiter	Successful
Venera 16	USSR	7 June 1983	Orbiter	Successful
Vega 1	USSR	15 Dec 1984	Lander and Balloon	Successful
Vega 2	USSR	21 Dec 1984	Lander and Balloon	Successful
Magellan	USSR	4 May 1989	Orbiter	Successful
Galileo	USA	18 Oct 1989	Venus flyby en route to Jupiter	Successful
Cassini	USA	15 Oct 1997	Venus flyby en route to Saturn	Successful
MESSENGER	USA	3 August 2004	2 Venus flybys en route to Mercury	Successful
Venus Express	Europe	9 Nov 2005	Orbiter	Successful
Planet-C (Akatsuki)	Japan	20 May 2010	Orbiter	Failed

Data from National Space Science Data Center, NASA Goddard Space Flight Center.

Appendix B
Data about Venus

Astronomical Data	Venus	Earth	Mars
Mean distance from Sun (10^8 kilometres)	1.082	1.496	2.2794
Comparative solar distances	0.723	1	1.524
Orbital period	0.615	1	1.881
Rotational period (hours)	5832.24	23.9345	24.6229
Comparative rotational periods	243	1	1.029
Comparative length of solar day	117	1	1.026
Comparative length of year	0.615	1	1.88
Orbital eccentricity	0.0068	0.0167	0.0934
Comparative eccentricities	0.412	1	5.471
Obliquity (deg)	177	23.45	23.98
Comparative obliquities	7.548	1	1.023
Equatorial radius (kilometres)	6052	6378	3397
Relative radius	0.95	1	0.53
Mass (10^{24} kg)	4.87	5.97	0.642
Relative mass	0.816	1	0.107
Mean density (kg/m^3)	5240	5500	3940
Relative density	0.950	1	0.714
Acceleration of gravity (m s^{-2})	8.89	9.79	3.79
Comparative surface gravity	0.877	1	0.379
Escape velocity	0.929	1	0.214
Solar Constant (kW m^{-2})	2.62	1.38	0.594
Bond albedo	0.76	0.4	0.15
Net heat input (kW m^{-2})	0.367	0.842	0.499
Molecular weight (g) *dry	43.44	28.98*	43.49
Specific heat C_p (J kg^{-1} K^{-1})	850	1005	830
Dry adiabatic lapse rate (K kilometres^{-1})	10.468	9.760	4.500
Surface temperature (K)	730	288	220
Surface pressure (atmospheres)	92	1	0.007
Mass (10^{16} kg)	4770	530	~1

Astronomical Data	Venus	Earth	Mars
Composition:			
Carbon dioxide	.96	.0003	.95
Nitrogen	.035	.770	.027
Argon	.00007	.0093	.016
Water vapour (variable)	~.0001	~.01	~.0003
Oxygen	~ 0	.21	.0013
Sulphur dioxide	150 parts per million	.0002 parts per million	~ 0
Carbon monoxide	40 parts per million	.12 parts per million	700 parts per million
Neon	5 parts per million	18 parts per million	2.5 parts per million
Krypton	<1 part per million	1.14 parts per million	0.3 parts per million
Xenon	<0.1 parts per million	0.087 parts per million	0.08 parts per million
Helium	~12 parts per million	5 parts per million	1 part per million

Index

Printed in the United States
By Bookmasters